CW00392812

COCK O'
THE NORTH

Gresley P2/1 Class 2-8-2 locomotive no. 2001 Cock o' the North *poses outside the Paint Shop at Doncaster Works for one of a number of official pictures taken after its completion in early May 1934.*

Fonthill Media Limited
Fonthill Media LLC
www.fonthillmedia.com
office@fonthillmedia.com

First published 2014

Copyright © Peter Tuffrey 2014

ISBN 978-1-78155-436-4

The right of Peter Tuffrey to be identified as the author of this work has been asserted by him in accordance with the Copyright, Designs and Patents Act 1988.

All rights reserved. No part of this publication may be reproduced, stored in a retrieval system or transmitted in any form or by any means, electronic, mechanical, photocopying, recording or otherwise, without prior permission in writing from Fonthill Media Limited.

Printed in England

COCK O' THE NORTH

GRESLEY'S BOLD EXPERIMENT

PETER TUFFREY

FONTHILL

A portrait of Herbert Nigel Gresley (knighted 1936), with stop watch, on a streamlined train c. 1936.

In August 1935 no. 2001 has been captured leaving Edinburgh Waverley station with the 9.55 a.m. passenger train to Aberdeen.

Contents

The book is dedicated to the late Malcolm Crawley

Acknowledgements

I am grateful for the assistance of the following people: Doug Brown, Miriam Burrell, Marian Crawley, Peter Jary, John Law, Hugh Parkin, Pat Plant, Derek Porter, Bill Reed, Staff at the National Railway Museum, David Rogerson, Peter Sargeson, Alan Sutton, Anthony Watson.

Special thanks are due to my son Tristram Tuffrey for his general help and advice.

Photographs

All photographs and drawings, unless otherwise stated, are from the collection of Ben Burrell, the former Doncaster Works official photographer, or the late Malcolm Crawley, former chairman of the Gresley Society and A1 Society vice chairman.

Every effort has been made to gain permission to use the photographs in this book. If you feel you have not been contacted please let me know: petertuffrey@rocketmail.com.

Information

I have taken reasonable steps to verify the accuracy of the information in this book but it may contain errors or omissions. Any information that may be of assistance to rectify any problems will be gratefully received. Please contact me by email petertuffrey@rocketmail.com or in writing Peter Tuffrey, 8 Wrightson Avenue, Warmsworth, Doncaster, South Yorkshire, DN4 9QL.

On the back cover:

Author Peter Tuffrey is pictured with the original *Cock o' the North* nameplate.

Introduction

A regimental march, a single malt whisky liquor and a motorbike road racing title all carry the name 'Cock o' the North'. But we have to look back almost 500 years and to George Gordon, fourth Earl of Huntly for the origin of the name and the reason for its application. He held uniquely powerful positions in Scotland during his lifetime making 'Cock o' the North' a fitting description of someone, or something, important and powerful ever since.

Hopefully, this gives some indication as to the reason why the name was chosen for the London & North Eastern Railway's new express passenger locomotive for services operating between Edinburgh and Aberdeen when it was completed at Doncaster Works in 1934. The P2 Class engine was the culmination of several years of design development undertaken by Gresley involving a number of L.N.E.R. classes, in addition to incorporating a variety of new features. His experience with the engine influenced later L.N.E.R. locomotive design. *Cock o' the North* was also the most powerful passenger locomotive to operate in Britain at the time.

Divided into three main chapters with sub-headings, this book features information and photographs which not only describe and illustrate Gresley's locomotive developments, but how these weaved a pathway to the building of L.N.E.R. no. 2001 *Cock o' the North*. The locomotive's construction and time in service will then be detailed before the final chapter deals with the circumstances surrounding no. 2001's rebuilding and final decade in service.

In the first chapter, a long step back is taken to the early years of the 20th century to focus firmly on Nigel Gresley after he had taken the reins of the Great Northern Railway Company's locomotive stock in August 1911. Gresley followed a great tradition of locomotive building at Doncaster and amongst his predecessors in the post were, Archibald Sturrock, Patrick Stirling and Henry Alfred Ivatt. Gresley did not hesitate to utilise some of their design practices, but he was bold enough to incorporate his own ideas into new construction at the works.

His first locomotive design, the G.N.R. Class H2 2-6-0 locomotive no. 1630, was built at Doncaster in 1912 and for this he recognised the various running practices adopted at home and abroad, particularly in the U.S.A., with regard to the wheel arrangement. The 'Mogul' 2-6-0 configuration had originated in America in the 1840s and was subsequently adopted by G.J. Churchward of the Great Western Railway shortly before being taken up by Gresley. The H2s were also the first class of locomotive in the U.K. to employ outside Walschaerts valve gear.

Over the next few years Gresley would move on to answer, in the short term, the G.N.R's freight and mixed traffic requirements. This left him free to explore possible applications of new features, such as the use of three cylinders. In order to do this, Gresley had to overcome the disadvantages of the application of this arrangement, namely the use of three sets of valve gear, by devising a system where the outside motion operated the valve for the inside cylinder. Gresley developed and obtained a patent for his conjugated valve gear allowing him to proceed with a three-cylinder development of his O1 2-8-0 Class and O2 Class locomotive no. 461 appeared in 1918.

At the same time, Gresley was developing a locomotive with a 4-6-2 or 'Pacific' wheel arrangement. Ideas for this design were probably galvanised by noting the introduction of the K4S Class Pacific (also their L1 Class 2-8-2 'Mikados' as later demonstrated) on the Pennsylvania Railroad in the U.S. The last great locomotive design produced for the G.N.R. was the A1 Class Pacific engine no. 1470 *Great Northern*.

At the Grouping of the railways in 1923, Gresley was not the first choice to be the London & North Eastern Railway's Chief Mechanical Engineer, but he found himself, aged 35, in the post and ready to take great strides. His first significant L.N.E.R designs were the powerful P1 Class 2-8-2 and U1 Class 2-8-0+0-8-2 locomotives, which appeared in time for the railway centenary procession in 1925, and they, along with the other Gresley designs present, left the railway world in no doubt of the direction of his design policy. Namely, this was the use of 'big engines' that were masters of their work because of large, free-steaming boilers that allowed the locomotives to operate within their capabilities with the trains of the day.

Gresley inherited a diverse collection of locomotive stock at Grouping and, despite producing a large number of his own designs subsequently, some of the former locomotives had to be retained because the L.N.E.R's financial position was never particularly stable or profitable throughout its existence. As a result, a number of classes were rebuilt with Gresley features to put them closer in performance to his own engines, or by using experimental equipment that aimed to increase efficiency. The main examples of this latter were the extensive trials conducted with feedwater heaters and poppet valves.

One man who found success with these features, in addition to his own developments of the locomotive steam circuit, was André Chapelon on the Chemins de fer de Paris à Orléans in France. Gresley took note of his thoughts on locomotive design and became well acquainted with Chapelon over the years.

The second chapter looks at the building of *Cock o' the North* and highlights that the engine incorporated a number of new features to the L.N.E.R. and ones that Gresley had considered and experimented with in previous years. These included: Kylälä-Chapelon (Kylchap) double blastpipe and chimney, a 2-8-2 'Mikado' wheel arrangement, Lentz poppet valves and rotary cam valve gear, A.C.F.I. feedwater heater, V-shaped cab front, streamlined steam passages and a 50 sq. ft firebox grate area.

Cock o' the North was constructed, principally, to work on the LNER's sleeping carriage traffic, which had been growing increasingly heavy during the summer months and was placing a strain on the locomotives assigned to the region. The route between the two cities was also particularly torturous, having an abundance of sharp curves, steep gradients and speed restrictions, requiring a powerful engine to keep to the scheduled times. Eight coupled wheels appealed to Gresley because they allowed good adhesion to the track for the transmission of the locomotive's power. The 'Mikado' wheel arrangement was particularly popular with foreign railways for similar routes.

Gresley was president of both the Institute of Mechanical and Locomotive Engineers and in his presidential addresses, and also in a number of technical papers, he appealed to his peers for a locomotive testing station to be erected in Britain. Unfortunately this did not happen during his lifetime, but early in December 1934 *Cock o' the North* travelled to Vitry-sur-Seine, near Paris, to undergo trials at the recently constructed locomotive testing station to obtain data about its performance. However, the locomotive encountered a number of problems at the station, predominantly hot axle boxes, and the tests were cut short.

Cock o' the North was joined in service by no. 2002 *Earl Marischal* in October 1934, which was significantly different from no. 2001 by having piston valves operated by Walschaerts/Gresley motion and an exhaust steam injector, and then by a further four - no. 2003 *Lord President*, no. 2004 *Mons Meg*, no. 2005 *Thane of Fife* and no. 2006 *Wolf of Badenoch* - in 1936. These latter engines were the same mechanically as no. 2002 because the poppet valves and

A.C.F.I. feedwater heater had not displayed their potential efficiencies in service. Therefore, they were removed from *Cock o' the North* in April 1938. At this time no. 2001 also acquired a streamlined front, which had been developed by Gresley for use on his A4 Class Pacifics and applied to the final four P2s when new.

Chapter three deals with the rebuilding of *Cock o' the North* after Gresley died in harness during 1941. Edward Thompson was made Chief Mechanical Engineer of the L.N.E.R. and a very different design policy pervaded the air of the Doncaster Drawing Office. Standardisation with a view to easing the maintenance burden was at the forefront of Thompson's thinking because of the Second World War and while he retained some Gresley features a number were discarded leading to the re-construction of some of the latter's engines. The most notable casualties of this course of action were no. 2001 *Cock o' the North*, the other P2 Class locomotives and no. 4470 *Great Northern*. The P2 Class had suffered from a number of mechanical and operating problems, which could have possibly been remedied if there had been desire to do so, but they gave Thompson the necessary excuse to rebuild the class as prototypes for his own mixed traffic A2 Class Pacifics. *Cock o' the North* was altered in September 1944, bringing to an end to a great period of innovation in locomotive design on the L.N.E.R.

The engine and other P2 Class rebuilds were not trouble-free in their new guise and were viewed with acrimony by footplatemen and enthusiasts alike, but they were capable of some good performances. *Cock o' the North* was scrapped in 1960.

Cock o' the North at King's Cross June 1934.

Chapter One Background and Development

Name Origin

The sobriquet 'Cock o' the North' has been in use for almost 500 years and has traditionally referred to the head of the Clan Gordon, beginning with George Gordon the fourth Earl of Huntly. He was born in 1514 to John Gordon, Lord Gordon, and Margaret Stewart, who was the illegitimate daughter of James IV of Scotland. George Gordon's father died in 1517 before the earldom could pass to him, so when Alexander Gordon, third Earl of Huntly, died in 1524 his estates passed to his grandson. As he was still in his minority Margaret Tudor, the Queen Mother and widow of James IV, was made his guardian and he grew up in the Scottish Court with King James V, who was only two years his senior.

The fourth Earl of Huntly was appointed Lieutenant General of the North when he was only 22 years of age and as part of this role he had to deal with the bitter feuds that arose between the Highland Clans. Then, when King James V travelled to France to form an alliance through marriage, George was appointed as one of the King's Regents in addition to Warden of the Marches. After the death of James V in 1542 the Earl of Huntly was again made a Regent, but this time it was for the infant Queen Mary and by 1544 Huntly had also become Lord Chancellor, presenting him with a uniquely powerful position in Scotland. The eleventh Marquess of Huntly, in his book *The Cock o' the North* (1935), relates that it was during this period that the name 'Cock o' the North' was first applied to the fourth Earl of Huntly: '...there is little doubt that it [the name] came into general use after this visit of the Regent Arran and Mary of Guise, the Queen Mother [of Mary, Queen of Scots] to Huntly Castle [whilst on their way to Inverness] where they were magnificently entertained by the Earl and his Countess, Elizabeth Keith.' The Marquess of Huntley adds: 'It is said that the Queen Mother, in conversation with one of her suite, remarked to him that their host was known as "The Cock o' the North." "Ah Madame," was the reply, "take care you do not have to clip the wings of this 'Cock o' the North.'" The story was repeated and it became current.'

While the Queen Mother did not have any need to take up such advice, the Earl of Huntly's power angered some Scottish nobles, especially Queen Mary's illegitimate half-brother Lord James Stewart, who, both publically and privately, did all he could to undermine Huntly and turn the Queen against him. As a result of this influence Huntly was stripped of the Earldom of Moray and the title was bestowed upon James Stewart. A number of unfortunate incidents then occurred which put Huntly in an unfavourable light and the Queen ordered his arrest and subsequent outlaw. In response Huntly gathered a small force and rode on Aberdeen, but he was met at Corrichie, a short distance away, by the Earls of Atholl and Moray at the head of a force that outnumbered his own. Despite assuming an early offensive, Huntly had to surrender, but the Earl died shortly after this of apoplexy. All of Huntly's titles, lands and possessions were then forfeited to the Queen and the Earl of Moray, as were those of his sons. Huntly's second son, also named George, became the fifth Earl after returning to favour in the subsequent years and all the family's titles were restored to him in 1565. In this year he was also made Lord Chancellor and became a leading figure in the Scottish politics of the day, following in the footsteps of the first 'Cock o' the North.'

The fifth Earl's son and sixth Earl was created Marquess of Huntly in 1599 and this title was held until the fourth Marquess of Huntly was created Duke of Gordon in 1684. The fifth Duke of Gordon and eighth Marquess of Huntly was the last of the male line and died in 1836, passing his estates to his nephew Charles Gordon-Lennox, the fifth Duke of Richmond. The sixth Duke of Richmond was subsequently made the Duke of Gordon in 1876 in the second creation. The title of Marquess of Huntly passed to George Gordon, fifth Earl of Aboyne and remains in this branch of the Gordon family.

George Gordon (1770-1836) became the fifth Duke of Gordon and eighth Marquess of Huntly and was the last 'Cock o' the North' from the original male line of succession.

Gresley Becomes Locomotive Engineer of the G.N.R.

After the announcement of H.A. Ivatt's intention to retire, the Great Northern Railway appointed H.N. Gresley to be his successor as the company's Locomotive Engineer on 8th August 1911. He was to take up the position on 1st October, having been with the G.N.R. for six years prior to this as the Carriage & Wagon Superintendent. Despite initial doubts as to Gresley's ability because of his age (35 at the time of the former appointment), he would acquit himself well, meeting the company's locomotive needs up to the time of the Grouping. Gresley produced a number

H.A. Ivatt (1851-1923), Locomotive Engineer of the G.N.R., 1896-1911.

Nigel Gresley, pictured in the early years of the 20th century.

G.N.R. Class H1 2-6-0 'Mogul' no. 1200 was constructed for the company by the Baldwin Locomotive Works in America and entered service during December 1900.

The first Gresley locomotive - G.N.R. H2 Class 2-6-0 no. 1630 - was a bold design from the new Locomotive Engineer.

O.V.S. Bulleid (1882-1970) was Gresley's 'right-hand man' from 1912 to 1937.

of new classes during this time, mainly for freight and secondary passenger services, that utilised the best features of his predecessor's designs, while incorporating a number of his own developments that would become the defining features of his later classes.

The first Gresley locomotive, G.N.R. Class H2 (L.N.E.R. K1) 2-6-0 no. 1630, was erected at Doncaster Works, not long after his appointment, in August 1912. The class were produced to negate the need to use passenger engines on goods services as there was a lack of suitable motive power at the time. The 2-6-0 or 'Mogul' wheel arrangement, which had originated in America in the early 1840s, was chosen because it had a number of advantages over the 0-6-0 wheel arrangement that would have been the alternative at the time. The addition of a pony truck would permit higher speeds to be run safer and steadier than an 0-6-0 and would also relieve the wear on the leading coupled wheels. A year prior to the introduction of Gresley's 2-6-0, G.J. Churchward of the Great Western Railway had designed a mixed traffic 2-6-0 and this had found success on the duties asked of it. There were some similarities between the G.W.R. and G.N.R. engines, both having 5 ft 8 in. diameter driving wheels, but there were differences between the boiler, firebox and cylinder dimensions. With regard to the motion, the H2's valves were operated by outside Walschaerts valve gear and the H2s were the first class of locomotives in Britain to employ it; on the G.N.R. Ivatt had used it previously, but only in a few isolated instances. Also, the Walschaerts gear was improved by Gresley to provide greater support for the radius block, which could be prone to slipping, and this was subsequently adopted elsewhere. In addition to the Walschaerts valve gear, Gresley introduced his patent double-bolster swing link suspension on the pony truck for the class and he subsequently applied this to all his designs that utilised this feature. In the suspension the upper and lower bolsters were connected by swing links

J.G. Robinson became Locomotive Superintendent of the G.C.R. in 1902 and held the position until 1922.

Ivatt Large Atlantic locomotive no. 1456 poses with the quintuplet carriage set built for the King's Cross-Leeds service in September 1921. The formation, consisting of a brake first, first class diner, kitchen car, third class diner and brake third, was mounted on articulated bogies and also featured all-electric cooking equipment. This was the first use of the method in the world and functioned through batteries connected to dynamos, or a mains connection at stations. The project was instigated by Gresley and overseen by Bulleid, who spent some time as Carriage and Wagon Superintendent for the G.N.R. and unofficially for the L.N.E.R. immediately after Grouping. The Leeds set was in service until 1953, but no. 1456 was condemned some time earlier in August 1947.

G.N.R. O1 Class 2-8-0 no. 456 was the first of five locomotives built to the design at Doncaster between December 1913 and March 1914. Fifteen were later built by the North British Locomotive Company.

and it was intended that the weight on the pony truck wheels would be equalised when the engine negotiated curves in the track. Ten H2s were produced before it was deemed that any further examples would require a larger boiler and subsequent locomotives built to the design, the G.N.R. H3 Class (L.N.E.R. K2), were so fitted; the ten H2s were also later rebuilt to conform.

In 1912, O.V.S. Bulleid returned to the G.N.R. as Personal Assistant to Gresley and the two men would develop a great affinity for each other over the ensuing years. Bulleid had previously been a premium apprentice at Doncaster under H.A. Ivatt and had begun the four-year course in January 1901. Upon completing the apprenticeship, Bulleid held Personal Assistant positions with Locomotive Running Superintendent Webster and F. Wintour, who was Doncaster Works Manager. Bulleid

spent just over a year in the latter position before he was lured to the continent for a post as Assistant Works Manager and Chief Draughtsman with the Westinghouse brake company in France. In 1910 Bulleid left this job as he was recommended to join the British Board of Trade by F. Wintour as Mechanical and Electrical Engineer of overseas exhibitions. Bulleid was employed for the 1910 exhibition in Brussels and Turin in 1911. At the end of the latter Bulleid enquired at Doncaster if there were any positions available and Gresley offered him either the District Locomotive Superintendent job at Grantham or a position as his assistant. Bulleid did not hesitate to accept the latter. The role allowed him to concern himself with all locomotive matters and kept him in close contact with the staff, especially those in the drawing office and he often acted as an intermediary between them and Gresley. Later,

G.N.R. Class J22 0-6-0 no. 546 was the first of the second order for the Gresley-designed members of the class to be completed at Doncaster in May 1913.

No. 157 was the pioneer J23 Class 0-6-0T, being completed in December 1913. The locomotive is pictured at Doncaster Works on 8th January 1914.

in 1920, Gresley promoted Bulleid to Assistant Carriage and Wagon Superintendent and the latter was involved in the introduction of the Leeds quintuple articulated set, which also boasted the world's first all-electric kitchen car.

The next Gresley design followed similar lines to the H2 and featured outside Walschaerts valve gear, large diameter cylinders and a leading pony truck. G.N.R. Class O1 (L.N.E.R. O1, later O3) 2-8-0 no. 456 was built at Doncaster in December 1913 to alleviate the pressure on the Ivatt 0-8-0 freight locomotives and to increase the train loadings permissible so that the double heading of freight trains would not be necessary. To achieve an increased haulage capacity Gresley used a large boiler, 5 ft 6 in. diameter with a 24 element superheater, which was also soon to be fitted to the H3 2-6-0s.

At the end of 1912 and 1913 respectively, Gresley completed two designs that covered the remaining need for short distance goods and shunting locomotives in his J22 0-6-0 (L.N.E.R. J6) and J23 0-6-0T (L.N.E.R. J50 and J51) class of engines. The J22 was a continuation of an Ivatt design with slight detail modifications. The J23 dispensed with the saddle tank previously used by Doncaster 0-6-0 designs for shunting work and was replaced by side tanks to increase the adhesion factor. This was necessary because of the nature of the track in the West Riding of Yorkshire, which was to be their sphere of operation. The design of the side tanks was unusual in that they were sloped at the front to increase the view forward for the crew and were also recessed in the bottom at the front so that access could be gained to the motion.

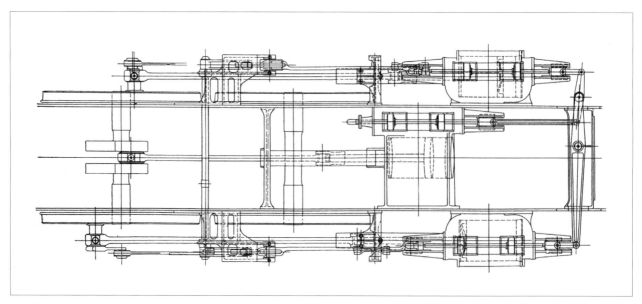

Gresley's conjugated motion. The valve spindles of the valves for the outside cylinders are extended forwards and connected to two levers; the equal lever and, the longer, '2 to 1' lever. These then transfer the motion to operate the inside valve.

No. 3279 (G.N.R. no. 279) seen in June 1938 after the replacement of its four cylinders with two measuring 20 in. by 26 in.

With the above designs Gresley had answered the G.N.R's freight and mixed traffic needs for the foreseeable future and with the passenger locomotive side being suitably provided for, through the use of the successful Ivatt 4-4-2 'Atlantic' and 4-4-0 engines, he was in a position where he could concentrate on developing designs. These would meet the future needs of the railway and explore possible applications of new features, such as the use of three cylinders. Up to the beginning of the 20th century two cylinders had mainly been in use on locomotives around the world. Three and four cylinders had been employed on a small scale, however, these applications were largely unsuccessful. In the early 1900s there was a greater demand for power because of an increase in train loads and three- and four-cylinder arrangements were

again attempted to harness more power for hauling the heavier trains. J.G. Robinson of the Great Central Railway produced a small number of three-cylinder engines and these were successful, however, an 'Atlantic' fitted with this arrangement was not. Vincent Raven (later Sir) of the North Eastern Railway had greater success with both three-cylinder freight and passenger locomotives. A number of four-cylinder engines were built in Britain by Churchward of the G.W.R. after he gained inspiration from work carried out in France, and these were also successful. Gresley gained first-hand experience with a four-cylinder locomotive through rebuilding Ivatt Atlantic no. 279 in May 1915. No information as to its performance is available, but it would appear that it was not such to warrant persisting with the arrangement.

Locomotive no. 365 Sir William Pollitt *was the last of four Robinson 8E Class three-cylinder compound engines to be built between December 1905 and December 1906 for the G.C.R.*

No. 483 of the second batch of O2 Class 2-8-0s, all of which were built in May 1921 by the N.B.L.C. These locomotives differed from no. 461 by having the simplified form of the Gresley conjugated valve gear (after its success on K3 no. 1000) and as a result the cylinder diameter was increased to 18½ in. Furthermore, the three cylinders' inclination was altered, the steam passages were improved, outside steam pipes were employed, the pony truck wheels were made smaller at 2 ft 8 in. diameter and the main frames were made slightly longer. No. 483 was in traffic until September 1963.

No. 1001 was built at Doncaster three months after the pioneer H4 2-6-0 had entered service in March 1920.

In a paper to the Institution of Mechanical Engineers on 'Three-Cylinder High-Pressure Locomotives' in 1925, Gresley expresses that he had seen the advantages of the three-cylinder locomotive for a number of years and goes on to list these as being: less coal consumption than with the two cylinder type of similar power; increased mileage between general repairs; less tyre wear than with the two cylinder type; reduced hammer blow through lighter reciprocating parts, therefore an increased axle loading would be possible; more uniform starting effort than two- or four-cylinder engines; lower permissible factor of adhesion and an earlier cut-off in full gear. Gresley states: '...if two instead of three main valve-gears could be used, as in the two- or four-cylinder engines, one of the principal objections which from time to time has been urged against three-cylinder engines would be removed.' By the end of 1915 Gresley had designed and applied for a patent for his valve gear for operating the valve for the inside cylinder and this was subsequently accepted in October 1916. As a result Gresley proceeded with a three-cylinder development of his O1 Class 2-8-0 and O2 Class no. 461 appeared in May 1918. After a series of successful trials ten more O2s were ordered and Gresley

announced that he would employ three cylinders in his future designs.

At the same time as he was developing his three-cylinder valve gear, Gresley was also developing a design for a locomotive with a 4-6-2 or 'Pacific' wheel arrangement to ultimately take the reins of the main line passenger traffic. The first outline drawing appeared in 1915 and resembled an Ivatt Atlantic locomotive. In the following year a feature on the Pennsylvania Railroad 'K4S' Class Pacifics appeared in *Engineering* magazine and it has been said by a number of writers that the design of the boiler used on this class influenced Gresley on his own Pacific boiler. The 'K4S' design was particularly noteworthy because it had been produced in quite a scientific manner. The boiler and firebox dimensions, in addition to the cylinder volume, had been developed through stationary tests on the Pennsylvania Railroad's Altoona test plant and corroborated by results obtained from tests in actual service. In conjunction with the 'K4S' Pacifics the Pennsylvania Railroad produced a 'L1' Class 2-8-2 'Mikado' locomotive which featured a large number of components with the 'Pacific', including the boiler. One feature of the 'K4S' design that was adopted by Gresley was the three-bar

The pinnacle of Gresley's design work for the G.N.R. - no. 1470 Great Northern.

No. 1470 Great Northern *in Doncaster Works' Erecting Shop, 24th March 1922.*

slidebar and it appeared on the majority of his outside cylinder locomotives; the first being the O2 Class of 1918.

In 1920 a new express goods engine, H4 Class 2-6-0 no. 1000 (L.N.E.R. K3), was built embodying the latest developments of Gresley design policy and these included a simplified version of his valve gear and a 6 ft 0 in. diameter boiler, which at the time was the largest in use on a British railway. After good performances from this locomotive in service the Pacific locomotive design was improved in light of the experience gained and the first two were ordered in January 1921. The first, A1 Class no. 1470 *Great Northern*, was completed at Doncaster in April 1922 and at the time Gresley stated that the aim for the new Pacific was to haul 600-ton trains at an average speed of 50 m.p.h., which was quite an increase in the average loads, by approximately 100 tons, hauled by the Ivatt Atlantic locomotives.

The second A1 Class Pacific, no. 1471 Sir Frederick Banbury, was built at Doncaster in July 1922. The locomotive is pictured at Doncaster station with an 'up' express shortly after entering service.

Gresley Appointed C.M.E. of L.N.E.R.

Gresley made quite an impression with the introduction of his two Pacific locomotives during 1922. However, the relative success of these engines, and indeed that of his other G.N.R. designs, was not enough to guarantee him the Chief Mechanical Engineer position in the London & North Eastern Railway when the company was formed

Gresley, around the time of Grouping.

on 1st January 1923. When he did get the position a few months later, locomotive designs from the constituent companies of the L.N.E.R. were perpetuated and it was not until 1925 that two Gresley L.N.E.R. designs appeared - the P1 Class 2-8-2s and the Beyer Garratt U1 Class 2-8-0+0-8-2. While these designs met a specific traffic requirement, Gresley subsequently produced a number of classes that met a more general need for either freight or passenger locomotives. These engines followed the principles Gresley had arrived at during his G.N.R. days, while a few classes featured innovations that attempted to bring efficiencies to the locomotive. An attempt was also made to make a number of the constituent companies' classes of engines more economical or to improve the design by bringing them closer to modern practices. The L.N.E.R. was subsequently badly hit by the trade depression in the early 1930s, as many other companies were, and this period is marked by the absence of any new locomotive designs, as well as very few built to existing arrangements. However, somewhat of a recovery was mounted in the middle of the decade and Gresley's locomotive design policy reached its zenith.

The L.N.E.R. was created through the Railways Act of 1921, also known as the Grouping Act, which brought together railway companies from specific areas of Britain to form four distinct companies; the London, Midland & Scottish Railway, the L.N.E.R., the Southern Railway and the G.W.R. The G.N.R., Great Eastern Railway, North Eastern Railway, Great Central Railway, North British Railway and Great North of Scotland Railway amalgamated to make up the L.N.E.R., but only three of the companies' Locomotive Engineers were in real contention for the corresponding position in the new company. They were Sir Vincent Raven of the N.E.R., J.G. Robinson of the G.C.R. and Gresley. Raven ruled himself out early in the race as he joined the board of Metropolitan Vickers, although he was also given an advisor role by the L.N.E.R. that

One of the first new L.N.E.R. designs - Beyer Garratt 2-8-0+0-8-2 U1 Class locomotive no. 2395 - completed in time for the railway centenary parade.

lasted until 1924 and during this time he produced two reports that were of benefit to the company. Robinson was well-qualified for the job, but his age was against him and he was ultimately given an advisor role before he retired. Gresley was thus announced as C.M.E. of the L.N.E.R. at a board meeting on 23rd February 1923.

Sir Vincent Raven, C.M.E. of N.E.R. 1910-1922, and briefly a contender for the C.M.E. position on the L.N.E.R.

The L.N.E.R. was not the largest railway company after Grouping, coming second to the L.M.S., and the financial outlook for the company seemed positive upon its formation. However, after the first year of operation the L.N.E.R. were faced with a £500,000 shortfall in the amount of money required to fulfil share obligations and interest payments, which was approximately £14.5 million, and the difference was consequently taken from the Government Compensation fund. This had been provided as part of the Railways Act as a means of compensating the railway companies for the Nationalisation of the network during the First World War. In the following two years the financial outlook became more and more bleak as in 1924 there was a strike of engine crews that led to a fall in traffic receipts and these dropped by a further £1.6 million in 1925 to provide a total of £58.2 million for the year. The general strike and miners' strike of 1926 contributed to a disastrous fall in traffic receipts to £48.6 million and the latter also added the additional cost to the company of importing coal. The Government Compensation fund was used further in 1924 and 1925 as £2.75 million and £4 million were withdrawn respectively. In 1926, £6.9 million was appropriated, in addition to funds from the company's contingency and general reserves to cover the

L.N.E.R's interest and share dividend payments. The situation improved to some extent in 1927 and then levelled off for the remainder of the decade as the net revenue remained in the region of £10 million, with an exception being 1929 when the figure rose to £12.3 million. Throughout the 1930s the traffic receipts of the company were predominantly in the region of £40 million, with the highest during this decade being £49.1 million, which was achieved in 1937. Traffic expenses had been slightly reduced from just over £50 million in the years 1923-1925 to £43.3 million in 1929. Then, during the 1930s the traffic expenses were reduced to roughly £35 million until the figure rose during the latter half of the decade. The net revenue of the L.N.E.R. was consequently poor during this period and after a high of £10.5 million in 1930 it reached a low of £7 million in 1932, but slowly recovered to £9.6 million in 1937. The worst year in the 1930s was 1938 when receipts were £46.6 million, traffic expenses were £40.5 million and the net revenue was only £6.1 million.

During the years of crisis the L.N.E.R. attempted to cut expenditure by increasing staff and locomotive efficiency, as in 1929 when a saving of £1 million in locomotive running costs was made as part of an effort by the footplatemen to be more economical in their driving methods. In addition, there were reductions in staff levels and wages, such as in 1932 when the company cut 5% from wages in an attempt to save £1 million, but this was later reinstated. Furthermore, in 1930 departmental expenditure was set strict limits and a section of staff was made responsible for making sure the restrictions in spending were adhered to. They were also tasked with locating areas where costs could be further reduced. There was also a sale of some of the L.N.E.R's assets and this included the sale of Tyne Dock in the late 1930s for over £800,000. Moreover, the L.N.E.R. took steps to reduce the amount of local authority rates they had to pay by lobbying the Government in the late 1920s and this was successful, reducing the rates bill by approximately £1 million. The company was also part of a drive by the four main railway companies to force the Government to relax regulatory restrictions on rate setting, but this campaign was subsequently interrupted by the start of the Second World War.

When Gresley was appointed C.M.E. he took the responsibility of the L.N.E.R's 7,423 locomotives, of 232 classes, that had been inherited at Grouping; there were 21,000 coaches and 300,000 wagons which also fell under his authority. The locomotives of the amalgamated companies varied in terms of design, power and age, with these attributes varying quite wildly within a company's locomotive stock. Furthermore, the First World War contributed to a general lack of new designs because the material and manpower had been needed elsewhere. The standards of maintenance had also suffered, with the resultant decline leaving some locomotives in a run-down condition and this situation was only just improving at the time of the Railways Act, which in itself made some companies reluctant to improve their rolling stock. There

More N.B.R. Class J 'Scott' 4-4-0s were asked for by the Scottish Area after Grouping, but were not forthcoming. B.R. no. 62419, N.B.R. no 410, Meg Dods is seen at work in the late 1940s.

was also a problem in that some of the locomotives were suited to the constituent's area of operation and the footplate crews could be wary of change and 'foreign' locomotives being introduced from a different area, as was demonstrated a number of times after Grouping.

During 1923 the L.N.E.R. constructed 126 locomotives to five of the six constituent companies' orders. The G.N.R. had ordered ten A1s and eight O2 Class engines; the G.C.R. eight B7 and ten A5s; the G.E.R. ten D16, ten J68 and three N7s; the N.E.R. thirty B16, ten J27 and five Y7s; the N.B.R. twenty-two N15s. In 1924 the engines from the 1923 orders were completed, while orders were placed for two Gresley

classes, A1 and K3 (40 and 50 respectively and to the new L.N.E.R. loading gauge to increase their sphere of activity), and a G.C.R. design. This was Robinson's 11F 'Director' Class and 24 of these 4-4-0s, classified D11/2, were constructed for use in the Scottish Area, which had actually made the plea to Gresley for more N.B.R. J Class, L.N.E.R. D29 and D30 'Scott' Class, 4-4-0s (it will be noted that the N.B.R. had virtually ceased all new construction when it was clear that the Railways Act would be passed). Gresley declined to authorise more of the latter class as he would shortly prepare his own 4-4-0 design incorporating his own principles. To meet the need short term the 'Director' Class was chosen because

Gresley's solution to the Scottish Area motive power crisis was to send 24 new D11/2 Class 4-4-0s which had a design of G.C.R. origin. No. 6389 Haystoun of Bucklaw *was one of twelve built by Kitson & Co. between July and October 1924.*

A view inside the smokebox of P1 Class no. 2394 showing the 'E double' superheater, taken in June 1926.

Initially Gresley had considered using a 2-10-2 wheel arrangement in conjunction with his Pacific boiler for the P1s, yet, this idea was discarded in favour of the 2-8-2 and the Drawing Office at Doncaster was ordered to proceed with the design in May 1923. Two locomotives were subsequently authorised at a meeting of the Locomotive Committee in August at an estimated cost of £8,000 per engine, but upon completion of both this had risen to over £10,000 each. The purpose of producing the P1 Class was for them to be utilised on large 100-wagon trains, weighing well over 1,000 tons, travelling between Peterborough New England and London Ferme Park marshalling yards. As an aid to achieving this both members of the class were fitted with a booster on their trailing axle. The booster drove the wheels on this axle through a small steam engine that could be engaged if the locomotive needed more power to start a train moving, especially when there were adverse weather conditions or difficult gradients. This piece of equipment was relatively new and had originated in North America only a few years prior to its application in the U.K. Gresley had obtained some experience of the booster before the installation on the P1s as Ivatt C1 Class 'Atlantic', no. 1419, L.N.E.R. no. 4419, was fitted with the component. However, in this instance its use was ultimately unsuccessful. Through the suggestion of the Superheater Company, who designed the booster, an 'E double' superheater was to be fitted to both engines, but in the event it was only fitted to the second engine, no. 2394, which was built at Doncaster in November 1925. The 'E double' superheater, with 62 elements and a heating surface of 1,104 sq. ft, was predicted to raise the steam temperature to 700°F., in comparison to the 575°F. achieved by the 32 element Robinson superheater fitted to no. 2393. In reality the predicted temperature was not achieved and through tests conducted with A1 'Pacific' no. 2562 *Isinglass*, which received the other 'E double' superheater originally to be used on no. 2393, the average temperature was only 31°F. higher than the Robinson superheater.

In service the P1s were more than capable of handling the heavy loads, but the main problem the class encountered was that the trains they were designed to handle were too long for the Operating Department to successfully find a route for. At Grouping the L.N.E.R. had intended to improve the track between Peterborough and London by providing longer refuge sidings and removing the bottlenecks, where the number of lines reduced from four to two. However, due to the financial position the company later found itself in, these improvements could not be implemented. The loads the P1s were then allocated had to be reduced from 100 wagons to 92 and then to 80 when the average speeds of passenger trains were increased. The latter figure was generally the load that a 2-8-0 would handle, thus invalidating the need for a P1 locomotive. The reduced train weights also meant that there was little use for the booster as the class had no trouble starting the lighter trains. Therefore, removal occurred in the late 1930s; maintenance problems also contributed to this decision.

it was the most powerful of the type that the L.N.E.R. inherited and these were augmented by Ivatt D1 Class 4-4-0s which were displaced from the G.N. section. The D11/2s were popular with the Scottish crews and worked between Edinburgh and Glasgow, Perth and Edinburgh, in addition to both north and south from Dundee on stopping and semi-fast trains. They could also be found at the head of, or double heading express services on the Edinburgh and Aberdeen line. The D1s were not as successful and were generally disliked in Scotland, being placed on menial duties or out of service.

In 1925 two new Gresley designs appeared in time for the centenary celebrations for the opening of the Stockton & Darlington Railway. Both were present in the procession of locomotives from all four of the railway companies (also attending for the L.N.E.R. were members of the A1, K3 and O1 Classes), in addition to some historic engines. The procession was one of the main attractions of the festivities and it was watched by the Duke of York, later to become King George VI, who was also attended by Gresley. The commemorations had been largely organised by the L.N.E.R., in particular A.C. Stamer (Assistant Chief Mechanical Engineer and Mechanical Engineer, Darlington) and Edward Thompson (Carriage and Wagon Engineer North Eastern Area), with Bulleid also being responsible for some of the details. One of the new Gresley designs present was P1 Class 2-8-2 'Mikado' no. 2393, which was completed at Doncaster Works at the end of June 1925, only just in time for the centenary procession as the celebrations began on the 1st July.

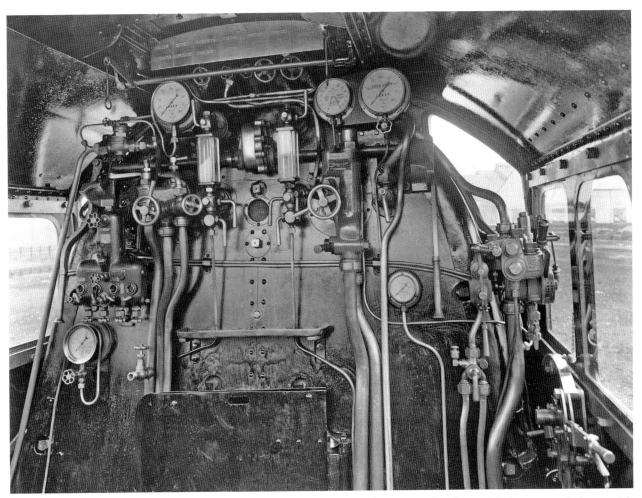

The cab of P1 no. 2393. It is interesting to note that the top three gauges have 'L.&N.E.R.' on their faces, while the gauge on the bottom right has 'G.N.R.'

P1 Class 2-8-2 Mikado locomotive no. 2393 poses near the Crimpsall Repair Shop at Doncaster Works.

Ivatt Large Atlantic no. 1419, equipped with a booster, 32 element superheater, piston valves and side-window cab in July 1923. The booster was present until November 1935.

Both of the P1 Class engines were later fitted with a diagram 94HP and a diagram 94A boiler respectively in the early 1940s, but they were withdrawn in July 1945 and became the first Gresley class to become extinct.

The other new design to appear in time for the centenary celebrations was the Beyer Garratt/Gresley U1 Class 2-8-0+0-8-2 locomotive, no. 2395, which was completed by Beyer, Peacock & Co. on 21st June 1925. The engine was constructed as a priority by the company so that it

The booster in position on no. 2393.

could take part in the procession and its frames had only been laid on the 1st of the month. The U1 had first been mentioned to the L.N.E.R. Locomotive Committee in October 1923 when it was the intention for the company to purchase two engines for the combined price of £20,000. By the April of the following year this had evolved to just one locomotive for the fee of £14,395 and the order was then placed with Beyer, Peacock & Co. At the end of July, Gresley changed the design so that instead of two cylinders being used at each end of the locomotive, there would be three cylinders with the valves being driven through the same motion used by the O2 Class 2-8-0s; this alteration added £500 to the cost. After the centenary parade no. 2395 entered traffic officially and became the first of its type to run on a British railway, in addition to being the most powerful due to its 7 ft 0 in. diameter boiler, six 18½ in. diameter by 26 in. stroke cylinders and tractive effort of 72,940 lb. No. 2395 was built for the specific task of banking coal trains, which emanated from Wath concentration yard, up the 1 in 40 Worsbrough incline and it was employed on this task until the Manchester to Wath line was electrified in the late 1940s. An attempt was made to employ no. 2395 on the 1 in 37 Lickey incline, on the route between Birmingham and Gloucester, in the 1950s, during which time the locomotive was also converted to burn oil. After these proved unsuccessful, the Beyer Garratt was condemned in December 1955.

Bulleid continued to be involved in carriage and wagon issues immediately after Grouping despite his official role in the new company being Principal Assistant to Gresley. Bullied was able to oversee the transition of carriage construction from the constituent companies designs to the new standardised coaching stock drawings that were introduced in 1924 and also to the application of Clearing House standards for wagon stock. During this busy period, Bullied was able to refurbish the Royal Train carriage set,

N.E.R. Raven Pacific no. 2400 City of Newcastle *was tested against A1 Class no. 1472 shortly after Grouping.*

which had been designed by Gresley in 1908. Then in 1925, he was brought closer to locomotive affairs and was involved with the P1 design. At this time he was also amongst a party of railway engineers that met with their French counterparts to discuss compounding and, although Sir Henry Fowler was interested in using the system on the L.M.S., both Bulleid and Gresley were against such an application on the L.N.E.R.

In the first few years of the L.N.E.R's existence, Gresley's 'Pacifics' underwent a number of trials, first against the Raven Pacifics, which were classified A2 by the L.N.E.R., and then alongside the G.W.R's 'Castle' Class 4-6-0s designed by Charles Collett. The Raven Pacifics had been built for the N.E.R. as a hasty reply to the introduction of Gresley's Pacifics and were an enlarged version of the N.E.R. Z Class Atlantic, L.N.E.R. Class C7. The trials between the two Pacific classes took place in mid-1923 and consisted of a number of runs with the dynamometer car between London King's Cross and Doncaster. While it was found that the boiler design of the A2s was on a par with that of the A1s, the design of the cylinders, motion and the wheelbase were not suitably adapted for the new class from the Z Class arrangement. During the trials the A2, no. 2400, had kept close to full boiler pressure and had a high superheat temperature, while the A1, no. 1472, struggled to get close to its working pressure of 180 psi and had a lower superheat temperature. The A1 Pacific was more economical, with regards fuel consumption, as it used 3.94 lb of coal per drawbar horsepower hour, whereas the A2 figure was slightly higher at 4.29 lb. These results were not enough to convince Gresley that the A2s would play anything more than a supporting role to his Pacifics and forty A1s were ordered in October 1923.

In anticipation of A1 Pacific no. 1472's attendance at the British Empire Exhibition, Wembley, in 1924, the engine was renumbered and named no. 4472 *Flying Scotsman*. Also present at this important event was the G.W.R's 'Castle' Class locomotive no. 4073 *Caerphilly Castle* and a prominent proclamation was displayed near it commenting that the engine was the most powerful passenger locomotive in operation in Britain. There is a degree of uncertainty as to how the subsequent trials between the two classes were instigated, nevertheless they took place at the end of April 1925 to determine how close the engines were in their performance. A1 no. 4474 and 'Castle' no. 4074 *Caldicot Castle* were pitted against each other on the G.W.R's line between Paddington and Plymouth, while A1 no. 4475, replaced by no. 2545 on day two, and 'Castle' no. 4079 *Pendennis Castle* ran against each other between King's Cross, Grantham and Doncaster. The results, on both the L.N.E.R. line and G.W.R., favoured the 'Castle' Class in terms of fuel consumption as the engines were 3.7 lb and 6 lb more economical of the coal used per mile on these lines respectively than the A1 Pacifics. The G.W.R. publicity department subsequently exploited the results and when the British Empire Exhibition was held later in 1925 no. 4079 *Pendennis Castle* was displayed next to, the returning, no. 4472 *Flying Scotsman* (it is also worth noting that the L.N.E.R. authorised eminent railway writer C.J. Allen to broadcast details of the L.N.E.R. runs on the BBC). Gresley was quite surprised at the coal consumption results and attributed it, specifically, to the higher boiler pressure employed by the G.W.R. engines. However, Gresley was of the opinion, at this time, that the savings made in fuel use, because of the higher pressure, would be offset by higher boiler maintenance costs.

No. 2545 was a late replacement for no. 4475 when the A1s were pitted against the G.W.R.'s 'Castle' Class.

While no. 4079 *Pendennis Castle* was present on L.N.E.R. territory for the trials, the opportunity presented itself for the valves to be taken apart and measurements taken. The 'Castle' Class used long travel valve gear and this had been a feature of G.W.R. engines since the early years of the 20th century after its introduction on the railway by Churchward. He had realised that the objections to having long travel valve gear were only pertinent when in relation to slide valves (principally in relation to wear) and upon the adoption of piston valves the disadvantages were negated. The use of long travel valve gear allowed a better flow of steam in and out of the cylinder, meaning the locomotive could be driven with shorter cut-offs and the regulator fully open, which improved the flow of steam between the boiler and cylinders. Gresley used short travel valve gear on the Pacifics and his other engines and the valve travel had been limited further since K3 locomotive no. 1000's centre valve cover had been damaged by the valve spindle crosshead when it had been running at a high speed. As a result Gresley was sceptical that any advantages could be gained from the application of long travel valve gear to the A1 Class. However, after the measurements had been

No. 4472 Flying Scotsman *was in immaculate condition while on display at the British Empire Exhibition, Wembley, as can be seen in this picture taken during March 1924.* Locomotion *can also seen on the left.*

An important development was the introduction of long travel valve gear. No. 2555 Centenary *was the first to be modified.*

procured from no. 4079 *Pendennis Castle* he authorised no. 4477 *Gay Crusader* to be fitted with long travel valve gear. Unfortunately, as applied this arrangement was not very successful and only strengthened Gresley's position against the idea. Bert Spencer, Gresley's Technical Assistant, who had been campaigning for his own long travel valve gear arrangement to be used on a Pacific, now found it even more difficult to have his proposal accepted. Gresley finally relented in October 1926 and Spencer set about developing the valves and valve gear to suit the Pacifics in much more detail than had been applied to the setting of the valves of no. 4477. The first A1 to be fitted with long travel valves was no. 2555 *Centenary* in March 1927 and it later showed an 11 lb saving in coal per mile over an A1 with short travel valve gear, with water consumption also reduced, and, in addition, an improved all-round performance. The order to convert all of the class was made in May 1927, but the task was not completed until 1931. A number of other

classes were altered to have long travel valve gear in order to increase the efficiency of the locomotive stock as a result of the savings produced in this instance.

In 1924, 35 J38 Class 0-6-0s were authorised for use on heavy goods and mineral trains in the Scottish area, and in particular the Fife coalfield. Although these locomotives were originally intended to be the first of an 0-6-0 Group Standard design, it was later deemed that larger diameter driving wheels should be employed, 4 ft 8 in. to 5 ft 2 in., to make them more generally suitable to the number of different areas under the L.N.E.R's control. The first members of the J39 Class were built late in 1926, only a few months after the completion of the J38 Class, and continued to be produced in batches until the early 1940s when the total in service reached 289. The design work for the J38 and J39 Classes was carried out at Darlington Drawing Office and it employed a number of features in the tradition of the works as well as those used at Gorton and Doncaster.

No. 1448 was the first engine of the largest Gresley Class - the J39 0-6-0s for goods traffic - and it was completed at Darlington in September 1926.

The acquisition of large numbers of ex-R.O.D. 2-8-0s and the original Robinson 8K class members curtailed the construction of Gresley's O2 Class. No. 5388 was built at Gorton in February 1914 for the G.C.R.

Gresley further boosted the L.N.E.R's freight locomotive stock through the acquisition of a number of ex-Railway Operating Division (Royal Engineers) 2-8-0s, which had been built for use in France during the First World War. They were built to the design of Robinson's G.C.R. 8K Class, with slight modifications, and a total of 521 were eventually constructed. A number of these went to the continent and others were loaned to a number pre-Grouping railway companies before being put into store with a view to being sold. The L.N.E.R. inherited 131 of the G.C.R. 8K Class at Grouping and by the end of 1923 a further 125 ROD engines had been bought at a price of £2,000 each. Another order for 48 was placed in 1925, reducing the number of J38s that would be constructed, but the price had now been cut to £1,500 per locomotive. A final order for 100 was placed in 1927 for the incredibly low price of £340. The G.C.R. and the R.O.D. engines formed the O4 Class and all were virtually a standardised design. Gresley attempted some modifications to the class from the late 1920s onwards, principally concerning the boiler, with a number of variants of the O2 Class diagram 2 boiler being tried.

During 1926, Gresley was introduced to the French locomotive engineer André Chapelon, who would later be behind some of the most important developments in locomotive design, while he was in England to visit Davey, Paxman & Co. Ltd, which held the patent to the Lentz poppet valve design. Chapelon was born in Saint-Paul-en-Cornillon, Loire, France, on 26th October 1892 and later studied at the École Centrale des Arts et Manufactures. His education was interrupted by the First World War, where he served for the French Artillery and, incidentally, he developed an improved form of artillery fire control for the French Army. Chapelon resumed his studies in 1919, competing them in 1921 and upon graduating as Ingénieur des Arts et Manufactures he obtained work with the Chemins de fer de Paris à Lyon et à la Méditerranée as a probationer in the Rolling Stock and Motive Power department. However, Chapelon was restricted by the

P.L.M. in the application of his ideas on thermodynamics and the scientific testing and development of locomotives, so he left the company and the railways to take up a position with the Société Industrielle des Téléphones. After a year, Chapelon found that he missed working with locomotives too much and through his thermodynamics professor at the École Centrale, Louis Lacoin, gained a recommendation to Maurice Lacoin, his cousin, who was the Engineer-in-Chief of the Chemins de fer de Paris à Orléans. In 1925 Chapelon took up a position on this railway with the Research and Development department and his first task was to improve the exhaust system of the railway's compound Pacific classes. Through tests it had been established that there were in a number of problems in the design of the steam circuit at the front end such as; throttling of steam at admission to the cylinders, high back pressure at the time of exhaust and an unacceptable reduction in the pressure of the steam in the intermediate receiver between the high and low pressure cylinders. To improve the operation of the locomotives Chapelon developed a new form of chimney and blastpipe, incorporating a device invented by Finnish engineer Kyösti Kylälä to improve the mixture of exhaust gases and steam, that created an even draught and expelled the well-mixed exhaust with as little effort as possible. In 1926 the Kylälä-Chapelon or 'Kylchap' exhaust system was applied to a number of P-O Pacifics and trials were carried out. The results were quite impressive as the back pressure was reduced, steam pressure in the intermediate receiver almost doubled, coal consumption was reduced and the steaming of the locomotives was considerably improved. Chapelon then set about redesigning the entire steam circuit of the Pacifics and the result of this would prove to be revolutionary.

Meanwhile, early in 1926, Gresley's three-cylinder 4-4-0 design was completed by Darlington. The first engine, no. 234 *Yorkshire*, appeared in October 1927 and was classified D49. The class were built to handle secondary passenger trains and to replace Atlantics where the need arose in the

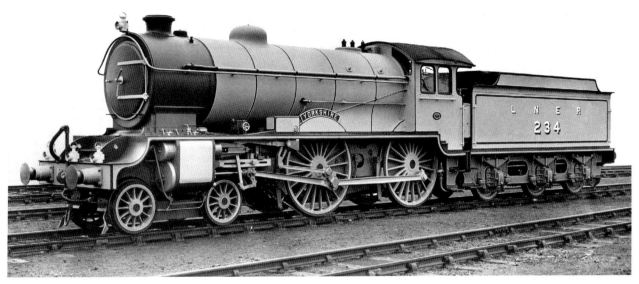

D49/1 Class 4-4-0 no. 234 Yorkshire.

North Eastern and Scottish areas. Twenty D49s appeared in 1927 and 1928 fitted with piston valves (class part one), these being operated by Walschaerts/Gresley valve gear. However, this arrangement was slightly different to that applied to other classes as the '2 to 1' lever was placed behind the cylinders instead of in front and Gresley's preference for the cylinders to drive on to the second coupled axle had to be dispensed with and all cylinders drove on to the first coupled axle. The cylinders followed Darlington practice and were cast in a single 'monobloc,' which was a feature adopted by Gresley in a number of later designs. Eight more D49s were ordered at the same time as the piston valve engines, but appeared after the last of these had been completed. The engines featured Lentz poppet valves operated by either rotary cam (class part two) or oscillating cam (class part three) valve gear. These locomotives will be dealt with separately.

An urgent need for a 4-6-0 on the Great Eastern section was brought to Gresley's attention in 1926. At this time the B12 Class (GER S69 designed by S.D. Holden) were under a great amount of strain due to the increasing loads on the main lines. Alleviating the situation with an existing design encountered difficulties because the former G.E. lines were subject to severe axle loading restrictions. There were also considerations to be made about the length of the locomotives and their adherence to the strict loading gauge. A number of K3s were brought into the G.E. section to bring some respite as this was only one of a few classes from the L.N.E.R. stock that was suitable. As a result, King's Cross Drawing Office was charged with producing an outline drawing for a new class of locomotive for the G.E. section and then Doncaster was asked to produce a detailed design to meet the specifications of; three cylinders, 25,000 lb

B17/1 Class 4-6-0 no. 2800 Sandringham *is pictured at Stratford shed and is carrying the 'Eastern Belle' headboard for the Pullman service between London Liverpool Street and Great Yarmouth.*

Third class sleeping carriage no. 1336 was the first of the ten diagram 148 carriages built at York in 1931.

tractive effort, 30 sq. ft grate area, 17 tons axle loading and D49 Class motion (piston valve engines). Doncaster drawing office struggled to meet these requirements and the North British Locomotive Company were then tasked with submitting a suitable design. In February 1928 the N.B.L.C. submitted two designs with axle loads of 19 and 18 tons respectively and with K3, O2 and A1 Class features as the company already had drawings of each class through producing a number of locomotives for all previously. The design with an axle load of 18 tons was ultimately chosen, which, it will be noted, was one ton over the specification and restricted the new locomotives area of work on the GE section. The N.B.L.C. also had difficulty fulfilling Gresley's requirement of all cylinders driving on to the second coupled axle and the drive had to be divided between the first and second axles; the '2 to 1' lever also had to be placed behind the cylinders as in the D49/1 Class engines. The first of the new locomotives, no. 2800 *Sandringham*, was classified B17 entered traffic in December 1928 and a total of 73 were constructed up to 1937.

The L.N.E.R. operated 56 sleeping carriages in 1923 and in the same year the company updated its sleeping car design to bring it into line with modern standards. Six vehicles were then constructed to the new specifications, diagram EC64B, at York, in addition to two sets of twin articulated sleeping carriages. A number of sleeping car services were provided by the L.N.E.R. to places such as; Newcastle, Edinburgh, Aberdeen, Inverness and London. However, these services were for first class and no third class berths were available. The L.N.E.R. did consider adapting six third class carriages for both day and night use, but these plans were later abandoned. Then, in May 1928, 16 convertible third class sleeping carriages were authorised and then built to diagram 95. Seven compartments were provided with seating for 56 in daytime and berths for 28 at night, as well as a lavatory at each end of the carriage. A pillow and rug were available for an additional charge of six shillings for journeys within England and seven shillings for those travelling to or from Scotland. The first services with the new carriages ran on 24 September 1928 and these included; the 7.30 p.m. King's Cross to Aberdeen and Inverness (one carriage for each

A1 Pacific no. 4480 Enterprise, *constructed at Doncaster in August 1923, was the first member of the class to receive the diagram 94HP boiler and was subsequently reclassified A3.*

No. 2544 Lemberg *was a participant in the A1 vs. A3 trials in 1928. The locomotive has been pictured during April 1932 as it is fitted with an automatic blowdown valve.*

destination), 9.15 p.m. St Pancras to Edinburgh and 10.45 p.m. King's Cross to Newcastle. The introduction of third class sleeping carriages was met with immediate success and by November a further twelve cars were authorised, with another seven being added in 1929. These were built to diagram 109 and were slightly different in having the width increased by 2¼ in. Ten more diagram 109 coaches were added to the building programme for 1930/1931 and the number of convertible third class sleeping carriages in use by the L.N.E.R. rose to 45. Also in the building programme were ten improved third class sleepers, diagram 148, and these comprised eight compartments, accommodating 32 berths, which were now fixed for night use only, and a washroom at each end of the car. The standard of the detail fittings was much improved and the berths were now provided with proper mattresses. The diagram 148 carriages found use on trains to and from Newcastle, Edinburgh, Aberdeen and Dundee. As a result of the introduction of third class sleeping carriages and subsequent improvement in their facilities, the first class accommodation was improved to stop custom being taken away from the latter.

While the long travel valve gear was being developed for the Pacifics, Gresley decided, after the 'E double' type superheater was deemed unsuccessful, that he wanted to increase the number of elements in the Robinson type already used by the A1s. A new boiler was designed early in 1927, diagram 94HP, to accommodate 43 elements, as opposed to 32 elements originally, with the increase being derived from a similar arrangement being used successfully by the German State Railway Co. The number of small tubes was reduced to 125, from 168, to increase the number of superheater elements and the heating surfaces then became 1,398.8 sq. ft in the small tubes and 706 sq. ft in the superheater. Shortly after the design for the 94HP began Gresley decided that he would abandon the relatively low pressure of 180 psi used by the A1 class and significantly increase this to 220 psi for the new boiler. At this time, it will be noted, the L.M.S. 'Royal Scot' Class 4-6-0s, which had just been ordered from the N.B.L.C., had a working boiler pressure of 250 psi, while the new S.R. 'Lord Nelson' Class 4-6-0s had a boiler pressure of 220 psi. The need for a higher boiler pressure for the Pacifics was also illustrated

No. 4473 Solario *represented the A1 Class in the 1928 assessment. The engine is photographed during November 1928 and this was to be used in conjunction with a picture of no. 2544 for the Doncaster Works Christmas card.*

A3 Class Pacific no. 2743 Felstead *entered traffic in August 1928 with corridor tender no. 5330.*

in a number of trials performed by the L.N.E.R., although not trials related to boiler pressure. The first instance was during a trial between Newcastle and Edinburgh when no. 2573 *Harvester* was at the head of a 520-ton train and had difficulty in getting up a steeply graded section of the line. Then, no. 2563 *William Whitelaw* was tested on the Waverley route, between Edinburgh and Carlisle, the locomotive encountered a similar problem, albeit with a number of other factors against a good performance. Five high pressure boilers were ordered from Doncaster

Works in March 1927, the first was fitted to no. 4480 *Enterprise* in July and the locomotive was reclassified A3 as a result. The next locomotive to receive a 94HP boiler was no. 2544 *Lemberg* in December and it was then chosen to participate in trials against an A1 with long travel valve gear; to make these more representative no. 2544's cylinders were lined to 18¼ in. diameter. The A1 chosen was no. 4473 *Solario* and the tests took place in February 1928 on the 11.04 a.m. train between Doncaster and King's Cross and returning on the 4.00 p.m., with each

W1 Class 4-6-4 no. 10000's boiler looking from the firebox end to the front. The boiler consisted of a steam drum (top), two front water drums and two at the rear, and a large number of tubes. The working pressure was 450 psi.

No. 10000 under construction at Yarrow & Co., Glasgow, during mid-1929. Originally, a standard A1 Class smokebox was fitted, but it was quickly replaced for the sloped smokebox to aid smoke deflection.

locomotive working this schedule for a week. No. 4473 was tested first and used an average of 38.8 lb of coal and 33.1 gallons of water per mile. The highest speed reached by the locomotive was 80.5 m.p.h. and the maximum drawbar horsepower produced was 1,128; it may be added that adverse weather conditions played a role in these results. No. 2544 was in charge of the trains, which were on average slightly heavier than those behind no. 4473, for the second week and the weather was slightly improved. The locomotive used an average of 35.3 lb of coal per mile and 30.2 gallons of water per mile. The highest speed reached by the locomotive was 83 m.p.h. and the maximum drawbar horsepower produced was 1,180. Both engines were worked with full regulator and low cut-offs, where practicable, and the drop in pressure between the boiler and cylinders for both engines was very low. Despite there not being much to choose between the engines ten new A3 Class locomotives, ordered in August 1927, were constructed at Doncaster between August 1928 and April 1929; erection of the first, no. 2743 *Felstead* began at the end of February 1928. Two more batches of eight and nine followed in 1930 and 1934/1935 respectively and the A1 Class, apart from no. 4470 *Great Northern*, were rebuilt to A3 specifications between 1939 and 1948.

With the introduction of the A3 Class in 1928 and the application of long travel valve gear to a growing number of A1s, an improvement of the average speeds of the mainline passenger services was made possible. However, this had been forbidden by an agreement made after the 1895 'Races to the North' (when the east coast and west coast railway companies challenged one another to see who could travel from London to Scotland in the fastest time) and the daytime schedule of 8 hours 15 minutes between London and Scotland had to be observed. In 1932 this was broken by the L.N.E.R. and L.M.S. and an acceleration of services ensued with the 'Flying Scotsman' service, which was made non-stop in 1928 through the introduction of the Gresley innovation of a corridor tender, being timed at 7 hours 30 minutes non-stop or 7 hours 50 minutes for the regular working. The reduction of train times was not just applied to the main line services, but were also applied to the Great Central section services between London Marylebone and Manchester and on the G.E. section between London Liverpool Street station and Norwich and Ipswich.

Since 1924 Gresley had been involved in designing a locomotive with a water-tube boiler working at a high pressure. Gresley had been attracted to this idea because of the economies in coal consumption this could offer and he was also partly inspired through the introduction of such a locomotive in America which had been built by the American Locomotive Company for the Delaware and Hudson Railroad. Yarrow & Co., shipbuilders, had been consulted for the design of the boiler, with which they had a good deal of experience for marine applications, and in September 1924 Gresley approached them in relation to designing a water-tube boiler for use on the L.N.E.R. The design process took approximately three years to complete and it was November 1927 before the boiler was ordered from Yarrow & Co. Then, throughout 1928 Darlington

No. 10000 is seen at Darlington Works in December 1929.

drawing office worked out the other details of the design. The boiler, which worked at 450 psi, was fitted to the frames at Yarrow's during 1929 and the engine, no. 10000, was completed at Darlington in November. The locomotive had a 4-6-4 or 'Baltic' wheel arrangement (although it has been claimed it is in fact a 4-6-2-2), four compound cylinders (it was intended at one point to employ Lentz poppet valves) and a corridor tender. No. 10000 did not enter traffic straight away after completion as it underwent a number of test runs both north and south of Darlington, in addition to a journey between Edinburgh and Perth, which turned out to be less than impressive. Modifications were made as a result of these trials, which included alterations to the superheater, injectors and blastpipe and the engine then entered normal service in June 1930. However, the locomotive returned to Darlington Works in August for further modifications and it was January 1931 before the engine was ready for service again. This pattern of trials, time in works for modification and brief spells in service was perpetuated for several years until Gresley finally decided

S.N.C.F. 231 E 6 was one of 20 Pacifics rebuilt by Chapelon and bought from the P-O by the Nord in the mid-1930s. 231 E 6 is seen in Gare du Nord during April 1958. Photograph courtesy of Bill Reed.

Another Pacific bought by the Nord was no. 3.1175, S.N.C.F. 231 E 5. The locomotive is 'on shed' at Calais on 30th August 1966. Photograph courtesy of Bill Reed.

to abandon the project. In October 1936 no. 10000 was sent to Doncaster to be rebuilt with a conventional boiler and it emerged in this form, also streamlined, in November 1937. The engine in its original form, it may be added, with its unusual shape, design and grey livery, was used extensively in its early years at exhibitions of locomotive and coaching stock by the L.N.E.R. and a short film of the locomotive after its first trip was also made by British Pathé News as part of their newsreel series shown in cinemas.

Just as no. 10000 was being completed at Darlington, Chapelon was ready to unveil his rebuilt 3500 Class Compound Pacific, no. 3566 (which was nicknamed *Cholera* by the crews, giving an indication as to its performance before the transformation), from the P-O's Tours Works. After improving the exhaust system for the company's Pacifics, Chapelon studied the engines' steam circuit and found that the cross-sectional area of the steam passages needed to be doubled; the steam chest volume had to be made four times greater (to eliminate throttling of the steam between the boiler and the cylinders), and sharp bends in the steam passages should be eliminated as far as possible. Chapelon also sought to improve the temperature of the steam in the superheater, as it was found that by the time the steam reached the low-pressure cylinders any heat gained from the superheater had been lost. To remedy this Chapelon sought to raise the temperature of the superheated steam entering the high-pressure cylinders by 100°C. These alterations, it was anticipated, would improve the locomotive's efficiency by 10%, while the steam circuit enhancements would offer 25% and the Kylchap double blastpipe and chimney approximately 20%. As a result Chapelon expected the locomotive to produce 3,000 indicated horsepower. In addition to the improvements in design, no. 3566 was also fitted with a Société l'Auxiliaire des Chemins de Fer et de l'Industrie (A.C.F.I.) feedwater

heater, Lentz poppet valves, operated by oscillating cam valve gear, and a Nicolson thermic syphon in the firebox. No. 3566 ran its first trial on 19 December 1929 and was 25% more economical at normal power outputs than it was originally, while between 75 and 80 m.p.h. 3,000 indicated horsepower was produced. On subsequent trials the locomotive's ability with heavy trains was remarkable and high average speeds could be maintained with ease, irrespective of gradients, and the service timings were also greatly improved upon. P-O Pacifics nos 3501-3520 were rebuilt in light of the improvements demonstrated by no. 3566 and formed the 3700 Class, with no. 3566 being renumbered 3701. As a further mark of the impact the alterations made, the Chemins de fer du Nord ordered 20 rebuilt Pacifics from the P-O as well as a further 28 new engines to the design from contractors. The Chemins de fer de l'État and the P.L.M. also employed some of Chapelon's principles in their designs.

In 1930 the 4500 Class Compound Pacifics of the P-O Railway were struggling with loads of up to 500 tons on the line to Toulouse, which was steeply graded in parts, and Chapelon was consulted for a solution. The first suggestion had been that the P-O's 2-8-2s should be rebuilt, but Chapelon countered that the Pacifics would be more suitable if they were rebuilt with a 4-8-0 wheel arrangement in addition to the alterations applied to Pacific no. 3566. The trapezoidal firebox carried by the 4500 Class was discarded in favour of a narrow firebox, which was quite long at 12 ft 6 in., and work began on converting no. 4521 in October 1931, this being completed in April 1932 at Tours Works. On the first run, at the head of a 642-ton train, no. 4521 surged up the steep banks and astonished all with a performance that was a further improvement on the rebuilt 3500 Class Pacifics. This was again demonstrated on another test, with a 575 ton train, between Vierzon and Limoges, when this difficult section of track was run at an average speed of 61.25 m.p.h. and no. 4521 produced sustained indicated horsepower figures of 3,800 and 4,000 at 56 and 70 m.p.h. respectively. In terms of drawbar horsepower, no. 4521 later renumbered no. 4701, continuously produced 3,030 at 62.5 m.p.h. and in 1935, after eleven more engines had been rebuilt to form the 4700 Class, no. 4707, with a train of 650 tons on a gradient of 1 in 143, produced 3,200 at 53 m.p.h. A number of times on the same trial the engine continuously produced figures of 2,750 drawbar horsepower and above.

L.N.E.R. no. 10000 and Gresley's V1 Class 2-6-4T engines were the last new designs to be produced for the company until 1934 as the global financial crash and subsequent trade depression seriously curtailed expenditure on new locomotive stock, with this also extending to carriage stock. In 1930, 74 new locomotives were built, but in the following year this was down to 69 and in 1932 it had halved to 34 engines. The worst year was 1933 when only 17 new locomotives were completed, comprising new B17, D49, J39 and O2 Class engines. For coaching stock, the 1930 production figure was 536 new vehicles, while in 1931 the number was 250. The worst year for carriage construction was 1932 when 65 new carriages were built by the L.N.E.R., but by 1933 the situation had recovered enough for 215 to be erected and in 1934 the figure was 379.

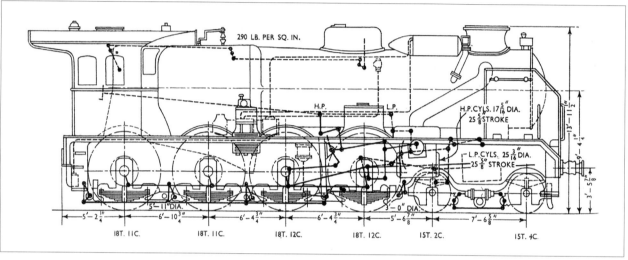

The diagram for Chapelon's rebuilt 4-8-0s which formed P-O Class 4700.

Four years elapsed between new Gresley designs after the appearance of the V1 Class 2-6-2T in 1930. No. 451 was constructed in October 1938 and is pictured during the same month.

Until the end of the 19th century the slide valve was the most common type of valve in use on railway locomotives around the world. However, a number of problems arose out of their use, such as; excessive wear of the valve and steam chest surfaces, difficulty in providing adequate lubrication and suitable lubricating oil and the valve events not being particularly accurate, which was becoming increasingly desirable at the time. The piston valve then became popular and its use spread rapidly because of its suitability to higher steam temperatures, which were becoming normal after the adoption of superheating, in addition to allowing more accurate valve events. With the piston valve the problem of wear remained, but this was not as marked as with the slide valve.

The poppet valve had been in use on stationary steam engines for a number of years and would also become extensively used in internal combustion engines. In the final years of the 19th century Dr Hugo Lenz (1859-1944) developed a poppet valve system for use in stationary steam engines, the first application being exhibited in Paris in 1900, and was later used by power stations in Austria and Italy. Dr Lenz then developed his poppet valve design for use on locomotives and this was later fitted, in the middle of the decade, by the Grand Duchy of Oldenburg State Railways to three of its P4.2 Class 4-4-0 locomotives. The use of the poppet valves then spread to a number of classes produced by the railway, but their application elsewhere was extremely limited. Also, at the time of this development, Davey, Paxman & Co. Ltd acquired the British patent rights for the Lenz stationary steam engine design and began to produce them in numbers, in addition to stylising the name as Lentz rather than Lenz. Then, in the early 1920s Davey, Paxman & Co. Ltd began to market and produce a Lentz valve arrangement for use in steam locomotives, both in Britain and abroad, with the patent rights being held by Lentz Patents Ltd, London.

Whereas slide and piston valves move over the steam and exhaust ports, poppet valves lift off their seats - there are two seats which are joined together - to allow steam in and exhaust out of the cylinder. As a result, the poppet valve gives the advantage of improved port openings, which allow a better flow of steam through the cylinder, and, furthermore, require little lubrication reducing problems of carbonisation of the lubricating oil. This makes the poppet valve attractive for use in conjunction with high temperature steam. For this reason the poppet valve was especially suited to locomotives working on the continent, where many of the railways employed high degrees of superheat as well as high boiler pressure with compounding, and did find extensive successful use there. The poppet valve can be placed either horizontally or vertically and the valves designed by Dr Lenz operate in the former plane, while, it may be noted, the Caprotti-type poppet valve work in the latter. Poppet valves can be moved by either oscillating cams, through the existing valve gear of the locomotive or rotating the camshaft continually. In the paper 'A New Infinitely Variable Poppet Valve Gear' by Miss V.W. Holmes, which appeared in the *Journal of the Institution of Locomotive Engineers*, no. 102, 1931, the most desirable features of a poppet valve gear are set out and it is worth recounting some of the points here. 'The cut-off should be infinitely variable, not limited to a series of steps. The lead [steam] should vary slightly, being greatest with early cut-offs, and least in full gear, in order to facilitate starting.' She adds: 'The valves should open and close rapidly, and should give a good area of opening even with early cut-offs. The cam box should be as compact as possible...be a standard unit, capable of rapid removal and replacement by a spare box.'

Gresley looked into the Lentz poppet valve design in 1923 and subsequently instructed that an 0-6-0 would be fitted with an oscillating cam arrangement. The engine chosen was no. 8280 of the J20 Class, formerly G.E.R. Class D81 designed by A.J. Hill and built in October 1922, and it entered traffic in May 1925 as the first locomotive in Britain to be fitted with poppet valves. As applied to

J20 Class 0-6-0 no. 8280 was the first locomotive in Britain to be equipped with Lentz poppet valves.

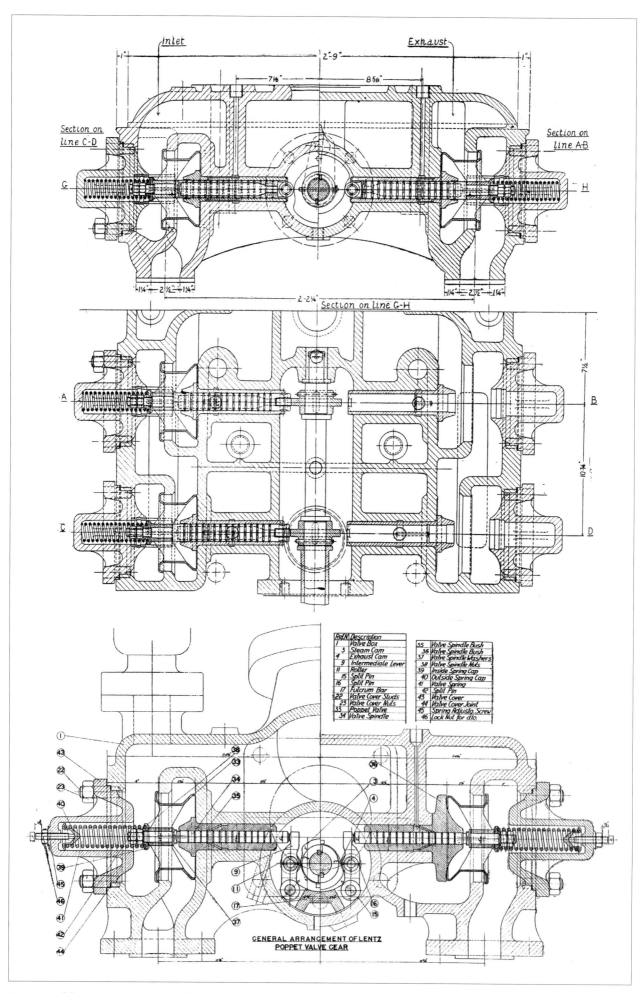

Section on line C-D

Section on line AB

Section on line G-H

Ref.N	Description		
1	Valve Box	35	Valve Spindle Bush
3	Steam Cam	36	Valve Spindle Bush
4	Exhaust Cam	37	Valve Spindle Washers
9	Intermediate Lever	38	Valve Spindle Nuts
11	Roller	39	Inside Spring Cap
15	Split Pin	40	Outside Spring Cap
16	Split Pin	41	Valve Spring
17	Fulcrum Bar	42	Split Pin
22	Valve Cover Studs	43	Valve Cover
23	Valve Cover Nuts	44	Valve Cover Joint
33	Poppet Valve	45	Spring Adjustg. Screw
34	Valve Spindle	46	Lock Nut for dto.

GENERAL ARRANGEMENT OF LENTZ
POPPET VALVE GEAR

Drawings of the valve chest as fitted to J20 no. 8280.

no. 8280 the Lentz poppet valves were housed in a valve chest that was bolted on to the two inside cylinders and the stepped cams were oscillated using the existing Stephenson motion. The poppet valves were also of the 'Double Beat' type, where the outer seating of the valve was purposefully made quite flexible, allowing the pressure of the steam to move the valve towards its seat and to keep the valve closed when required. The valve's two seatings, which took the form of lips, were made to slightly different diameters, these being, in this instance, 6⁹⁄₁₆ in. for the outer seating and 6³⁄₁₆ in. for the inner seating, both seatings being ⅛ in. thick. The lip of the outer seating was curved backwards to the face of the valve, while both seatings were machined to make contact with a raised ring on the valve chest casting which formed the valve seat. This feature was important as it allowed perfect contact for both seats, despite inequalities arising due to variations in expansion, and made the valves steam tight for long periods.

Bulleid, in his paper 'Poppet Valves on Locomotives', read to the Institution of Locomotive Engineers in February 1929, gives the details of the trials conducted with no. 8280 and J20 no. 8287, which was fitted with piston valves, between August and September 1926. No. 8280 used 67 lb of coal per train mile and 65.9 lb per engine mile, while no. 8287's figures were 70.3 lb and 69.3 lb respectively (it may be added here that no. 8287 was especially repaired for the trials and particular attention was paid to the condition of the piston valves). Bulleid states: 'These figures confirm the theoretical deduction that oscillating Lentz valves driven by ordinary motion will show a slight economy when compared with well-designed piston valves. The saving is due to the poppet valves remaining steam tight, and to the fact that the engine can be worked at rather earlier cut-offs.' Bullied also relates that no. 8280 had run a higher mileage between general repairs, 64,410, when compared with the average of the class, 46,642, and

he adds that the wear of the Lentz valve components was negligible within this mileage. No. 8280 ran with Lentz valve equipment until September 1937, when it reverted back to the J20 piston valve arrangement.

After no. 8280 had proved the Lentz valve arrangement was acceptable, Gresley made the decision to fit a 4-6-0 passenger locomotive with the feature. The class selected was L.N.E.R. B12, another former G.E.R. Class, S69, which had been introduced by S.D. Holden, and the engine was no. 8516. The Lentz valves were fitted in December 1926 and differed slightly from no. 8280's valves in having the valve chest cast in partnership with the two inside cylinders. The existing Stephenson motion was used to operate the valves, which were 6⅞ in. and 6½ in. diameter for steam inlet and 7¼ in. and 6⅞ in. diameter for the exhaust. Bulleid (1929) again gives particulars of the results from comparative trials with the locomotive, which took place during March and April 1927, and no. 8509. No. 8516 was favoured by a coal economy of 9.14% on an engine-mile basis, 9.14% on a train-mile basis and 11.17% on a ton-mile basis. In trials conducted during the following month between no. 8516, and no. 8518 the former was again favoured, but the advantage was not as marked, with the figures, in terms of coal economy, being 1.91% on an engine-mile basis, 1.76% on a train-mile basis and 3.4% on a ton-mile basis. In comparison with the rest of the class, no. 8516 used 53.7 lb of coal per mile, whereas the B12s with piston valves used, on average, 55.5 lb per mile. In terms of maintenance no. 8516 ran 68,386 miles between general repairs, while a B12 Class engine underwent a general repair every 57,657 miles, on average.

B12 no. 8525 was fitted with Lentz poppet valves in September 1928 and subsequently nos 8519, 8540, 8533 and 8532 were also altered to conform between July 1929 and April 1930, the locomotives now forming B12 Class

B12 Class 4-6-0 no. 8533 was one of several class members that were fitted with Lentz poppet valves. The engine had the equipment between September 1929 and February 1933.

D49/3 no. 318 Cambridgeshire.

part two. In addition, ten new B12/2 engines were erected in 1928 by Beyer, Peacock & Co. to boost the motive power available on the G.E. section. The Lentz valves and gear was slightly altered on these locomotives from that fitted to no. 8516. While the running of all B12/2 engines was quite satisfactory, particularly in terms of acceleration and free running at high speeds, a number of problems began to arise. These included cracks in the cylinders, which meant the costly replacement of the whole unit, and distorted camshafts. As a result, the step was taken to replace the Lentz valves with the standard

piston valve arrangement for the class, with long travel valve gear, and this was undertaken between November 1931 and February 1933. The locomotives that had been converted to the Lentz pattern valves had them removed between June 1932 and February 1934.

In mid-1926 Gresley gave instructions to Darlington Works for six engines to be fitted with Lentz valves operated by oscillating cams and the order for the equipment was placed with Lentz Patents Ltd in August 1926. By the following April the design for the valves had been completed, this being similar to those in use on no. 8280,

D49/3 no. 320 Warwickshire.

D49/3 no. 329 Inverness-shire *was the only member of the class with oscillating cams to be sent to work in Scotland.*

and the valve chest was again secured to the cylinder by a number of bolts. The design of the valves also coincided with Spencer's work on long travel valve gear for the A1 Class Pacifics and the design of the cams and the rocking lever arm were altered as a result. Walschaerts motion was employed for the outside cylinders, while the principle of operating the inside cylinder through the two outside valve gears was used, but differed from other applications to Gresley classes by having the motion levers arranged vertically. The first engine to be completed was no. 318 *Cambridgeshire*, which was constructed in May 1928 at Darlington, and it was followed into traffic by nos 320, 322, 327, 335 and 329 between May and August to form D49 class part three. From the outset, the locomotives encountered problems with the valve rods for the centre cylinder. The cause of this trouble was attributed to the double-eye coupling, between the valve rods and the camshaft, becoming overstressed as a result of the valve connecting rod coming into contact with the coupling when the engine was running in full gear. The valve travel for the centre cylinder was meant to be 6 in., but, in reality, there was 7⅝ in. valve travel and this contributed to the problem. To solve the troubles the double-eye coupling was given extra clearance and the maximum cut-off was reduced by 2½% to 62½%. Then, in June 1929, no. 329 *Inverness-shire* was fitted with the standard Gresley motion arrangement and, as a significant improvement in the locomotive's operation occurred, the other five D49/3 engines were altered between February and September 1930. Cylinder renewals were required by the locomotives after approximately ten years in service and Gresley decided at this time to remove the poppet valves and replace them with piston valves. This modification was carried out between March and November 1938 and the engines were then reclassified D49/1.

While the oscillating cam valve gear engines had seen some success, Gresley felt that the true advantages of the Lentz poppet valve system would not be seen until a successful rotary cam valve gear arrangement could be developed. He asked Lentz Patents Ltd to produce a suitable design that could be used on two locomotives of the D49 Class and he received and approved an outline design early in 1929. Detailed design work on providing the drive for the camshaft and a suitable reversing arrangement became drawn-out, but both were eventually produced. To make sure the former design was acceptable, no. 8280 was fitted with a prototype driveshaft and gearbox for the rotary camshaft and ran just over 4,500 miles with it fitted, developing no problems and running satisfactorily. As a result the design was adopted and two D49 Class engines, no. 352 *Leicestershire* (later *The Meynell*) and no. 336 *Buckinghamshire* (later *The Quorn*), were equipped and entered traffic on 13 March and 7 June 1929 respectively. The rotary motion for the camshaft was taken from a return crank located on the first pair of coupled wheels to bevel gears and then to a driveshaft which turned a further set of bevel gears connected to the camshaft. This latter was made of high tensile steel. The cams featured a number of different profiles to suit the desired valve events. The steam admission cams consisted of a series of rings, split into two distinct sections to allow forward and reverse running. In forward gear five rates of cut-off were provided; 15%, 25%, 35%, 50% and 75%. In reverse only two admission selections were provided; 75% and 35%. The exhaust cams were arranged so that when the engine was in full gear the point of compression started later than when a short cut-off was used. A cam was provided to hold the exhaust valves open when the engine was in mid-gear. Also present was a cylindrical cam to close the steam inlet valves when mid-gear was selected. An oil bath surrounded the whole mechanism to keep it well lubricated and free from detritus.

In service the two locomotives performed well enough for 15 more to be authorised in 1929 and these were erected between 1932 and 1933. Of these engines, no. 282 *The Hurworth*, built in October 1932, was fitted with a unique camshaft that allowed an infinite number of cut-off positions between 15% and 84%. However, the camshaft was removed after only 18 months in service and replaced with a camshaft that allowed seven cut-off positions to be

C7/2 Atlantic no. 732 has been pictured in the mid-1940s, perhaps shortly before its withdrawal in December 1946.

selected and these were; 15%, 20%, 25%, 35%, 45%, 60% and 78%. This type of camshaft was subsequently fitted to 25 new D49/2 engines built between 1934 and 1935; these locomotives also had slightly larger diameter valves.

B. Spencer in his paper 'The Development of L.N.E.R. Locomotive Design 1923-1941' recounts in some detail the results of trials carried out between the three D49 Class variants in December 1929. The engines concerned were; D49/1 no. 236 *Lancashire,* D49/2 no. 352 *Leicestershire* and D49/3 no. 329 *Inverness-shire* and all the trains hauled by the locomotives travelled between York and Newcastle. The average coal consumption per drawbar horsepower was, respectively; 3.68 lb, 3.62 lb and 3.69 lb. The average coal consumption per mile was; 35.1 lb, 31.55 lb and 34.66 lb. The average water consumption per drawbar horsepower hour was; 27.15 lb, 28.17 lb and 27.58 lb. The average water consumption per mile amounted to; 258.5 lb, 245.1 lb and 258.8 lb. The average drawbar horsepower exerted by the engines was; 430, 395 and 420. Spencer points out that no. 352 did less work than the other two locomotives and this would have influenced its figures. Another round of tests were conducted between the D49/1 and D49/2 engines in late 1935 and early 1936, using the L.N.E.R's counter-pressure brake locomotive, further suggested little difference between the piston valve and poppet valve systems. The piston valve locomotive, no. 251 *Derbyshire* was more economical when running normally, while the performance of the poppet valve engines, D49/2s no. 292 *The Southwold* and no. 377 *The Tynedale*, highlighted the need for a reduction in the cylinder clearance volume. The D49/2 Class locomotives were the only locomotives built by the L.N.E.R. to retain their Lentz poppet valves for the whole of their careers and the engines were withdrawn between December 1957 and March 1960.

Also playing a supporting role in the experiments with Lentz poppet valves and rotary cam valve gear were C7 Class Atlantics nos 732 and no. 2212. No. 732 was rebuilt during 1933 with 17 in. diameter by 26 in. stroke cylinders and 6⁵⁄₁₆ in. diameter exhaust valves, with the cams allowing five cut-off positions. Amongst the other alterations to the locomotive at this time were the fitting of Woodard connecting rods, which were also seen on a number of D49 Class engines, a new crank axle and a new pattern bogie from the one originally fitted. As a result of the changes carried out the locomotive formed C7 class part two. No. 2212 had been built at Darlington in June 1918 as part of the N.E.R's foray into the improvement of steam distribution, being fitted with Stumpf 'Uniflow' cylinders. This arrangement appears to have been satisfactory, but did not warrant further exploration. During the 1930s when the cylinders required renewal a decision was taken for the engine to be equipped with modified D49/2 Lentz valves with seven cut-off positions. No. 2212 re-entered service in this form during January 1936. The result of these two rebuilds would appear to be that it offered no appreciable difference over the standard piston valves fitted to the other C7s and no other class members were similarly altered. Nos 732 and 2212 retained the Lentz valves until they were withdrawn in December 1946 and October 1945 respectively.

A few comments regarding the number of cut-off positions that were available to the driver are worth noting here. *Locomotives of the L.N.E.R. Part 3A* by the Railway Correspondence and Travel Society gives a brief illustration of no. 732 and its performance. Early in September 1936, A4 Class Pacific no. 2510 *Quicksilver* had to be taken off the 'Silver Jubilee' service after running hot and was substituted at York for no. 732. The locomotive ran as far as Doncaster where it was replaced at the request of the driver, from the GN section, by an Ivatt C1 Class Atlantic as he could not gain the required performance from the engine because of the limits imposed by the five cut-off positions. In *The Railway Magazine* of November 1941 the driver of no. 732 is reported as commenting in similarly negative terms to O.S. Nock that: 'one cut-off position was not sufficient to do the job, and the next, if used, would "just about have killed the fireman", to quote the driver's own words.' This sentiment is also expressed by Mr R.J. Thackeray, Shed Superintendent, Neville Hill Shed, Leeds, when commenting on the lecture, given by Mr E.C. Poultney, on 'Poppet Valves as Applied to Locomotives' and a summary of which appears in the *Journal of the Institution of Locomotive Engineers* (1930). Mr Thackeray remarks, in relation to the D49/2 Class: 'In the selection of the cams for the rotary valve gear there is room for improvement, as when lifting heavy loads from stations and ascending gradients, the step between full gear and second position is too great.' However, he adds: 'At the same time, when the engines fitted with rotary cam valve gear have a good run, they attain high speed and work very smoothly...up to the present the fuel consumption of the poppet valve engines in our service compares favourably with that of those fitted with the piston valves.'

Gresley did not limit his experiments with poppet valves to the Lentz design as he also sought experience of the Caprotti arrangement. The G.C.R. Robinson 4-6-0 Class 9P, L.N.E.R. B3, locomotives were a small number of four-cylinder express passenger engines built between 1917 and 1920 and had a reputation for consuming excessive amounts of coal. In December and September 1929 respectively, no. 6166 *Earl Haig* and no. 6168 *Lord Stuart of Wortley* had Caprotti poppet valves and rotary cam valve gear fitted in an attempt to reduce the coal consumption. The Caprotti arrangement consisted of four valves per cylinder, one for inlet steam and one for exhaust at each end, and these were operated by a similar

rotary cam design to that fitted to the D49/2s. The distinct difference between the two applications, however, was that the cut-off positions for the Caprotti valves were infinitely variable. In service the coal consumption of the Caprotti B3s was reduced by an average of 16% in relation to the average of the piston valve B3s. After overcoming a few teething problems with the arrangement, the application was extended to no. 6167 *Lloyd George* in June 1938 and no. 6164 *Earl Beatty* in June 1939. The operation of the valves on the latter was slightly changed so that they were re-seated using steam instead of springs. These four B3 Class engines were the sole Caprotti application on the L.N.E.R. and the locomotives retained the valves until withdrawal, which occurred between September 1946 and December 1947.

Both the G.W.R. and the L.M.S. also applied Lentz poppet valves to their own designs of locomotive, with the former fitting the arrangement, operated by rotary cams for nine cut-off positions, to 29XX 'Saint' Class 4-6-0 no. 2935 *Caynham Court* during May 1931. This was the only locomotive altered by the G.W.R. and it carried the valves through to withdrawal in December 1948. The L.M.S. attained more experience with both the Lentz and Caprotti valves and the latter were first fitted to former London and North Western Railway four-cylinder 'Claughton' Class 4-6-0 no. 5908 *Alfred Fletcher* in 1926. As on the L.N.E.R., the application here was to effect a reduction in the coal consumption of the engine, which was extremely high in this instance, and it garnered a sufficient improvement for nine other 'Claughtons' to be converted.

After the emergence of the first D49/2s on the L.N.E.R., the L.M.S. decided to test the Lentz poppet valves with rotary cam valve gear on its Hughes 'Horwich Mogul' or 'Crab' Class 2-6-0s. Five engines were converted, nos 13118, 13122, 13124, 13125 and 13129, between December 1931 and February 1932. The method of operating the valves was the same as on the D49/2 engines, but the cams only provided four cut-off positions and three

Robinson 9P Class, L.N.E.R. B3, 4-6-0 no. 1168 Lord Stuart of Wortley *as built for the G.C.R. in October 1920.*

B3 Class no. 6164 Earl Beatty *was modified to incorporate Caprotti valves and rotary cam valve gear. The engine is pictured after the cam box covers were removed completely as regular overheating of the equipment was occurring.*

in reverse. The steam inlet valves were 6⁹⁄₁₆ in. diameter and the exhaust valves were 8 in. Subsequently, as a result of tests against a piston valve member of the class, it was concluded that there was no appreciable difference between the two varieties. Although the poppet valve engine had considerably reduced back pressure at the cylinders, there was a marked pressure drop between the boiler and cylinders in comparison with the piston valve locomotive. In 1933 further comparison tests were carried out on freight trains and again the difference between the two types of engine was only slight. A problem on the engine fitted with Lentz valves that arose during these trials was a large amount of spark throwing at high cut-off positions, but this was alleviated to a certain degree by reducing the cut-off. Some tearing of the fire also occurred because of the quick release of the inlet valves. Despite these results all five of the Hughes 2-6-0s kept their poppet valves until withdrawal.

Mr E.S. Cox, who for a number of years had been employed by the L.M.S. and later became Executive Officer (Design) for British Railways, made a number of comments in the discussion of Spencer's 1947 paper. Amongst these were a number of observations about the application of poppet valves on the L.M.S. He commented: '...so far as dynamometer car tests were concerned, the results obtained [by the L.N.E.R.] were much the same as were obtained by the L.M.S., namely, that under test conditions there seemed very little to choose between them in coal consumption.' He adds: 'It was a pity that tests were not continued to obtain coal consumptions over a shopping period, so as to obtain data as to the steam-tightness of the poppet valve; because he [Cox] thought that it was there, rather than in any special efficiency when new, that the great advantage of the poppet valve might be expected to arise.' Cox then asks Spencer if there are any comparative maintenance cost figures for the piston valve and poppet valve D49s. Spencer replies that there are not, but comments: 'L.N.E.R. experience certainly shows that the poppet valve engines are the cheaper to maintain. There were D.49 Class engines in service to-day with the original cams and valves.'

Feedwater Heater Experiments

The feedwater heater was developed as a means of recovering some of the heat energy present in exhaust steam, which would have otherwise been wasted, and using it to heat water before it was introduced into the boiler. The use of such an appliance, therefore, would effect a reduction in fuel consumed by the locomotive fitted. There would also be a reduction in maintenance costs, as introducing hot water instead of cold water into the boiler reduced metal stresses and any repairs needed as a result. Numerous locomotive engineers around the world have attempted to capture the potential rewards from the employment of a feedwater heater and many different types have been developed and available for use. These have met with varying degrees of success.

The feedwater heater can either be of the 'open' type that mixes the exhaust steam and the feedwater together to raise the temperature of the water or the closed type, which mixes the two separately, through suitable heat exchange apparatus, and the exhaust steam is then rejected. The former type has the advantage of being the simpler of the two and reducing water consumption through the retention of the exhaust steam. A disadvantage is that a suitable and effective oil separator is needed to remove any lubricating oil present in the exhaust and stop it from entering the boiler. Furthermore, a pump has to be used to transfer the hot water into the boiler and this has either been operated through a mechanism coupled to the movement of the engine's wheels or by steam.

Professor Edouard Sauvage, in his paper 'Feed-Water Heaters for Locomotives' (1922), explains the benefits, in terms of efficiency, of heating the boiler feedwater: 'Utilization of the exhaust steam is, in principle, a very simple process, and its efficiency is easy to calculate. For instance, on a locomotive, with an effective steam-pressure of 14 kg. per sq. cm. (200 lb. per sq. in.), and superheated to 340°C. (644°F.), the temperature of the feed-water being 15°C. (59°F.), the production of 1 kg. (2.205 lb.) of steam requires 723 calories. By heating the water from 15°C. to 95°C. (203°F.), that is using 80 calories, or which it may be assumed that three-fourths are given by the exhaust steam, sixty calories out of 723 are recuperated, that is 8.3 per cent., and the proportion is greater when the steam is not superheated. The service of the engine remaining the same as before feed-water heating, the fuel economy may be in excess of that figure, owing to increased utilization of heat in the boiler, due to a reduced production, the temperature of the gases rejected in the smoke-box being lower and the quantity of cinders smaller. On the other hand, with the same rate of combustion 9 per cent. more steam will be produced, 1 kg. of steam requiring 663 calories, instead of 723.' He adds: 'This calculation assumes that all the water passes through the heater, and does not apply to special cases where the nature of the service requires the frequent use of the ordinary injector.'

An early example of a feedwater heater was the Kirchweger arrangement. In this system exhaust steam was taken to the tender, condensed in the water tank to heat the water and the heated feedwater was then pumped into the boiler. This system was employed during the mid-1800s and found particular favour in Germany. John Chester Craven, locomotive engineer of the London, Brighton and South Coast Railway, employed a similar system on two 2-2-2 6 ft 6 in. 'Singles' built at Brighton in 1861. Over 50 years later the company's K Class of 1911 employed a similar method until it was removed in the late 1940s to early 1950s. A number of other feedwater heaters were developed from the 1850s to the beginning of the 20th century and these included; the Chiazzari pump, Lencauchez heater and Copeland heater. These operated on the 'open' principle, while the latter design differentiated itself by being having its apparatus located inside the smokebox.

During the early years of the 20th century a number of new designs were developed and brought into use by railways around the globe. Some of the principal designs in use were; Worthington, Weir, Knorr and A.C.F.I. feedwater heaters. In Britain the exhaust steam injector, predominantly of the Davies & Metcalfe or Gresham & Craven varieties, was the main piece of apparatus employed for re-using exhaust steam to introduce hot water into the boiler. The injector was usually used in conjunction with a live steam injector. The exhaust steam injector worked by taking exhaust, after it had been filtered by an oil separator, and passing it through a number of nozzles, introducing it to the feedwater at the same time. Once mixed the feed entered into the boiler through a clack valve. Live steam was also available to the exhaust steam injector.

Sir Vincent Raven, in the discussion of Sauvage's paper, notes that he had gained experience with the Weir arrangement and had come to the decision that he preferred the Davies & Metcalfe apparatus, but in conjunction with a combination injector. Raven expressed doubts, however, that the injectors were not without their own faults. He commented: 'The question of economy was also dependant, to a large extent, upon the human element, because the driver could either use the apparatus or not, as he pleased.' Raven adds: 'Again it was also dependant upon the perfection of the oil separator because if the separator was not absolutely clean, oil and impurities might get back into the boiler which would do infinitely more harm than if the exhaust injector were not used.' Raven also suggested that the economy figures worked out on paper did not always happen in practice. He noted that through trials on the N.E.R. with feedwater heating an economy of 5% was attained.

Gresley was made the first Chairman of Committee for the Leeds centre of the Institution of Locomotive Engineers when the centre was founded in 1918 and on this occasion he addressed the meeting of members. In his speech, Gresley made a number of points about locomotive design of the past and the necessary features of locomotives to be constructed in the future. Amongst the points for the latter issue, Gresley highlighted the need for a suitable feedwater heater if the efficiency of the locomotive boiler was to be improved. He reasoned that this was because engines were reaching the upper limit of weight and size which they could be built and therefore any improvements would have to be made within these constraints. In relation to feedwater heaters, Gresley commented: 'Many arrangements have been applied, but none have yielded such striking results as to justify their general adoption. In most cases, owing to the heaters getting blocked up, their efficiency is so much reduced that there is no return for the extra cost of fitting and maintenance.' He adds: 'The economy to be obtained by the introduction of a really satisfactory feedwater heater is second only to the economy which has resulted from superheating.' Further, Gresley observes that for perfect results to be obtained then the feedwater must be heated to the temperature of the water already in the boiler, but, he notes, this is not possible with exhaust steam heaters or a waste gas heater, this latter being generally precluded because of weight considerations. Gresley remarks that a live steam heater offers the most promising method of heating the feedwater to the desired temperature. He said: 'Live steam heaters have proved in marine work to be economical because to feed a boiler with practically boiling water increases the boiler efficiency. Although there is extracted from the live steam as much heat as absorbed in the feedwater, on the other hand the amount of heat transferred from the furnace to the water is increased.'

The first Gresley locomotives to carry a feedwater heater were nos 456-460 of the O1 Class. These engines carried a Weir heater and pump from new (December 1913 - March 1914), in addition to a live steam injector. In the Weir system the feedwater was pumped from the tender into a heater that used exhaust steam from the cylinders,

The Worthington heater mounted on no. 476.

as well as the exhaust from the pump, and the hot water was then discharged into the boiler on the left-hand side. No. 459 was subsequently altered in 1916 by having the heater removed and cold water was then fed into the boiler. Neither of these applications seem to have been a success as the equipment was removed from the locomotives between October 1918 and February 1920.

At the end of 1922, O1 no. 476 was equipped with a Worthington feedwater heater. This functioned along similar lines to the Weir equipment, but was mounted

O1 Class 2-8-0 no. 476 was equipped with a Worthington feedwater heater in November 1922 and carried the apparatus until mid-1927, a year after this picture was taken in June 1926.

externally on the locomotive's running plate. *Locomotives of the L.N.E.R. Part 6B* (1991) notes that no. 476 was tested with the apparatus between Ferme Park and Peterborough in 1923 and that 'the pump was easy to operate and the feed was easily maintained.' In 1926 two more applications occurred with B12 no. 8509 and C11 (N.B.R. Class H) Atlantic no. 9903 *Cock o' the North* the recipients. However, no. 476 had the Worthington device removed in mid-1927 and no. 8509 was relieved of it in 1929. No. 9903 kept it until withdrawal in May 1937.

The J6 Class of 0-6-0s (G.N.R. J22) also featured in Gresley's experiments. No. 522 was the first of the class to receive a top-feed arrangement in April 1916. This part consisted of a water chamber that was situated in front of the steam dome on top of the boiler. The chamber was filled to a fixed level with feedwater and then overflowed on to a trap that slightly heated the water; impurities present in the feed were also removed at this stage. The hot water was then fed into the boiler. Three more J6s were fitted before no. 522 was modified to carry an extra water chamber on the running plate to further remove impurities as they were causing problems with the original design. The water was also heated before it entered this water chamber through being mixed with live steam. These J6s carried the top-feed for a number of years, but it was taken off the engines between 1925 and 1931.

Gresley followed up his G.N.R. feedwater heating experiments with the first of two Dabeg feedwater heater applications after Grouping. The first was to O2 Class 2-8-0 no. 3500 in January 1925. The Dabeg apparatus functioned in the same manner as the Weir and Worthington heaters, but the pump was operated from the crank pin of the last pair of coupled wheels. The second fitting occurred in March 1926 when C7 Atlantic no. 2163 acquired the Dabeg system and it was a fixture on the locomotive for

11 years before being removed. No. 3500 retained it for several more years before the use of this feedwater heater on the L.N.E.R. ceased.

The A.C.F.I. type was the most extensively applied feedwater heater to the L.N.E.R's locomotives by Gresley. O.S. Nock, in *The Locomotives of Sir Nigel Gresley* (1945), suggests that there were up to 2,000 locomotives working on the continent fitted with the A.C.F.I. apparatus and savings of 10-12% in coal and 15% in water could be expected. In December 1927 L.N.E.R. Class B12 4-6-0s nos 8505, 8517 and 8523 received the arrangement, which was of the 'open' type. A suction reservoir, situated below the water level of the tender, delivered the feedwater to a tandem pump (with two cylinders for dealing with hot and cold water separately and operated by live steam) mounted on the running plate. From the pump the water was delivered to a water drum mounted on top of the boiler between the chimney and the dome. Here the water mixed with exhaust steam siphoned off from the blastpipe, the used steam from the water pump was also utilised, and both had passed through an oil separator before entering the mixing chamber. The water entered the drum and into a perforated pipe which forced the water to hit the top of the chamber and fall as a fine spray. The water was further divided by apertures in a metal plate and this allowed the water to be heated quickly. The product of the mixing chamber was then delivered into a second drum, located adjacent to the first, where gases produced by the process of heating were removed through a vent and liberated into the atmosphere. From the second chamber the water flowed to the hot water cylinder of the tandem pump and it was then admitted to the boiler.

C7 Atlantics nos 2206 and 728 were then fitted with A.C.F.I. heaters in January and February 1928 respectively.

O2/2 Class 2-8-0 no. 3500 was the first L.N.E.R. locomotive to receive a Dabeg feedwater heater. Note the connection to the rear coupled wheel crank pin which powered the equipment.

B12 no. 8523 was one of three engines of the class that acquired an A.C.F.I. feedwater in December 1927.

Locomotives of the L.N.E.R. Part 2A (1978) mentions that a report on how the equipment was functioning on no. 2206 was produced at the end of the year. The former states: 'The Darlington inspector responsible for the tests [pointed] out that the apparatus was inconvenient to use because of the number of valves that had to be operated, also that it had to be drained during frosty weather if the engine was dead.'

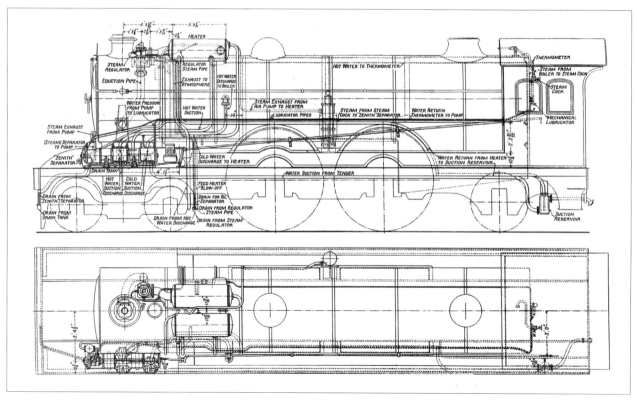

A diagram showing the arrangement of the A.C.F.I. feedwater heater as fitted to the two C7 Class Atlantics.

No. 2206 is seen at York station not long after being fitted with the A.C.F.I. apparatus.

A1 Class Pacific no. 2576 *The White Knight* and A3 Class Pacific no. 2580 *Shotover* received an improved A.C.F.I. arrangement during August and July 1929 respectively. In this instance the two water chambers were brought into the smokebox, to reduce heat losses through radiation, and were fixed to the engines in front of the chimney. Another important difference from the earlier application was that the heating of the feedwater in the mixing chamber by the exhaust steam was performed at the pressure of the steam rather than at atmospheric pressure and it was hoped that this would further increase the temperature

of the feedwater. The mixing chamber was located above the hot water chamber, with the former being the bulge at the front of the smokebox. *Locomotives of the L.N.E.R. Part 2A* (1978) also notes that during tests with one of the Pacifics fitted the temperature of the feedwater entering the boiler was 225°F. (107°C.).

Two years elapsed before the decision was taken to equip more members of the B12 Class and from 1931 until 1933 fifty locomotives were dealt with. The A.C.F.I. was not the same as on the A1s, but was the arrangement as fitted to the original B12 application, with slight alterations;

A1 Pacific no. 2576 The White Knight *has also been pictured at York, but the location is the locomotive depot.*

none of these engines were from the B12/2 Class part. Whilst this was taking place, the G.E. section had seen improvements that allowed a higher axle load to be accepted and steps were then taken to fit a new boiler to the B12 Class. The rebuilt locomotives were classified B12/3 and those undergoing the transformation equipped with the A.C.F.I. heaters did not retain them. Approximately half of the engines fitted were affected in this way, while the remainder had it removed under directive from November 1937. The A1 and A3 Class engines also fell under this command and it was stripped from these locomotives in December 1938 and February 1939 respectively. The two C7s were not dealt with until a specific command was issued in the early 1940s to deal with these engines

and the dates of removal are June 1941 for no. 2206 and March 1942 for no. 728.

In addition to the problems outlined previously, the A.C.F.I. suffered from chronic problems with scale formation, especially where the water was hard, and this necessitated acid cleaning as often as every six weeks. *Locomotives of the L.N.E.R. Part 2A* (1978) suggests 'the long term capital and maintenance costs did not justify the actual savings in coal and water.' In relation to the A1s it is added: 'There were a number of failures of the equipment often caused by enginemen untrained in the use of it. This arose because the engines were not kept to specific jobs and therefore were manned by men from other sheds.'

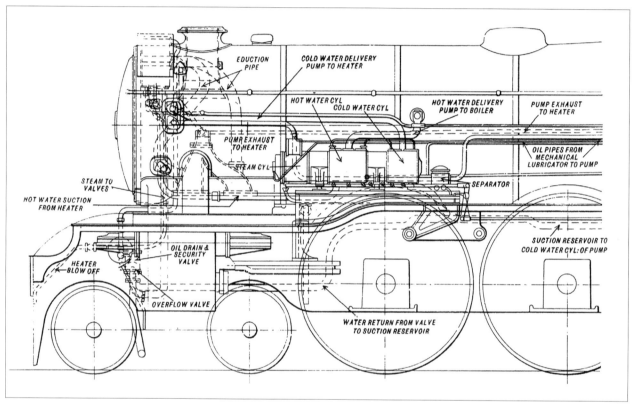

The arrangement of the A.C.F.I. equipment was the same for the A1 and A3 Class Pacifics.

A3 Class no. 2580 Shotover.

Chapter Two Construction and In Service

The Route

Since the mid-1810s there had been attempts to connect the cities and towns on the north-east side of Scotland to Edinburgh in the south by means of a railway. These had been made primarily for trade reasons, such as connecting the Fife coalfield with the rest of the country to distribute its product. Between 1817 and 1827 Robert Stevenson & Son surveyed a number of possible routes between the population centres of the north and Edinburgh and some of these were little altered when subsequently adopted by the relevant railway companies in later years.

Robert Stevenson was again in the area to survey a line between Edinburgh and Dundee in 1836, but his route lost out to a proposal submitted by Grainger & Miller. The partners, who were heavily involved in the engineering of railway lines in Scotland, produced a similar route to that of Stevenson, which progressed from Burntisland to Kirkcaldy, Cupar and then terminated at Ferryport-on-Craig, later known as Tayport. However, it was not until 1840 that enough support for the route was gathered to allow an Edinburgh, Dundee and Northern Railway Bill to be submitted to parliament. Both in 1842 and 1843 the Bill was defeated and this forced the project to be re-stylised as the Edinburgh & Northern Railway. A line from Burntisland to Cupar, with a branch to Perth, was authorised in 1845 and an extension to Leuchars and Ferryport-on-Craig was also later passed by parliament for the railway. Construction of the route began in early 1846, with 18 contracts for varying lengths of line being awarded to local builders. Progress became protracted because of land purchasing problems and labour issues. Despite this, the first section to Cupar opened to the public on 20th September 1847 and to Ferryport-on-Craig on 17th May 1848. The line had cost £744,000 and was £144,000 over budget. Ferries from Granton, Edinburgh, to Burntisland and Ferryport-on-Craig to Broughty Ferry conveyed passengers across the Firth of Forth and Firth of Tay respectively. Trains from Broughty Ferry, on the Dundee side of the Tay, were taken over Dundee & Arbroath Railway metals. The E.&N.R. became the Edinburgh, Perth & Dundee Railway in 1849 and was taken over by the North British Railway in 1862.

The Dundee & Arbroath Railway had been an earlier venture that had also utilised the services of Grainger & Miller to survey a line between the two places. The subscriptions for the line drew heavily from local landowners and business people and the Act for the line was passed in mid-1836. Eight contracts were subsequently awarded for the construction of the route's sixteen-and-a-half miles and the opening occurred on 6th October 1838. In 1845 the line, originally 5 ft 6 in. gauge, was brought into conformity with standard gauge and also at this time an agreement was reached with the Arbroath & Forfar

Railway (another Grainger & Miller line) and the Aberdeen Railway Company for running powers over their tracks, allowing Dundee and Aberdeen to be connected by rail.

Stevenson had also surveyed a line to Aberdeen in the mid-1820s, but it was not until 1845 that an Act was passed for a line to be built by the Aberdeen Railway Company from Aberdeen to Guthrie on the A.&F.R. The route took five years to complete after a number of delays due to company's poor finances, but rail travel between Aberdeen and Edinburgh, albeit with two ferry crossings, was now a reality.

The Aberdeen Railway Co. and the A.&F.R. ultimately became part of the Caledonian Railway, while the D.&A.R. remained independent for a number of years. However the latter subsequently became the Dundee & Arbroath Joint Line in 1879 as the C.R. and N.B.R. entered into a partnership to run the line. It was especially important for the N.B.R. as it offered the only access to their Arbroath to Montrose line, which was built between 1880 and 1883, and from the latter place to Kinnaber Junction for the line to Aberdeen.

The N.B.R. sought to improve the journey between Edinburgh and Dundee from the 1860s by removing the need to cross the two Firths by ferry as this was an obvious disadvantage for passengers or goods using the line. The problem was not solved until the late 1880s when the Tay was bridged successfully after the first bridge had collapsed. The Forth Bridge opened in 1890 and finally afforded a continuous rail connection between Edinburgh and Aberdeen.

The line between Edinburgh and Aberdeen (in relation to the period from the early 20th century to the 1930s) featured many undesirable characteristics that hindered locomotives from operating services to their optimum levels. The route had numerous adverse gradients (many around 1 in 100), moderate to severe speed restrictions on or near the gradients and badly sited stations, which were either on curves or at the bottom of a rising gradient.

From Edinburgh Waverley station the first steep gradient for a locomotive travelling on the line was a 1 in 100 up to Dalmeny and this was then followed by a 40 m.p.h. speed restriction over the Forth Bridge. The line fell to Inverkeithing station, where there was also a 25 m.p.h. service slack and next came one of the most severe rises on the route, a 1 in 94½ to Dalgetty. A quite steep descent then occurred to Burntisland where a top speed of 25 m.p.h. had to be adhered to. From here the track rose quite steadily to Kirkcaldy and there were then 3¼ miles of track that consisted of gradients varying between 1 in 143, 1 in 105, 1 in 100 and 1 in 114 to take the line to Dysart. Next came Thornton Junction, which was reached after traversing a declivity, and it was subject to a severe

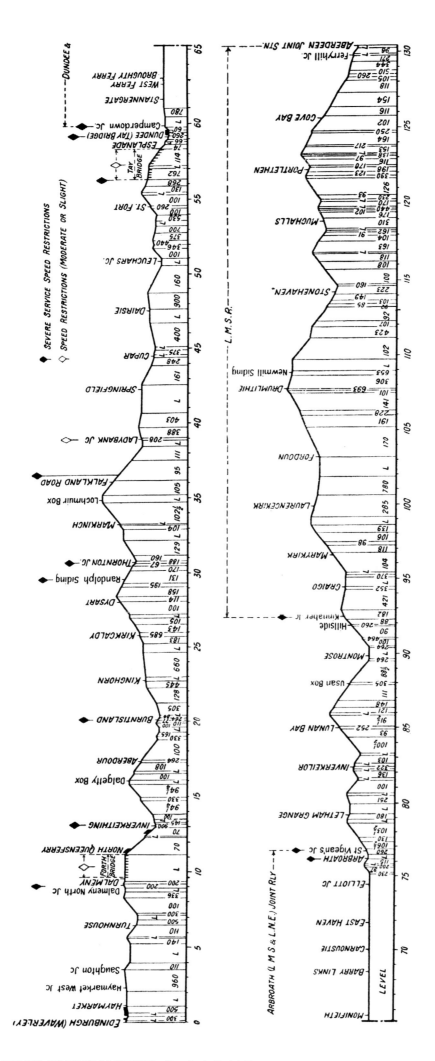

The gradient profile of the route between Edinburgh and Aberdeen.

15 m.p.h. speed limit as a result of mining subsidence. Lochmuir box was the highest point the track reached between Edinburgh and Dundee and it followed on from Thornton Junction. After climbing a 1 in 129 gradient, there was a level section to Markinch then approximately two miles of 1 in 102½ to the summit. The remaining 20 miles to Dundee were generally downhill, but there was a 1 in 100 ascent from Leuchars Junction and another two climbs of 1 in 100 before and after St Fort.

Almost immediately after leaving Dundee Tay Bridge station there was a short section of 1 in 60 to Camperdown Junction before a prolonged level section of track for roughly 15 miles. Then, a brief section of 1 in 97 had to be tackled before Arbroath could be entered. After Arbroath came St Vigean's Junction, where a 25 m.p.h. limit had to be observed, and from there the line rose in increments of 1 in 106½, 1 in 130 and 1 in 103½ to Letham Grange. Inverkeilor was passed before the next major acclivity had to be surmounted. The gradients on this section were 1 in 100½, 1 in 93, 1 in 252 and 1 in 91½ to the peak before a descent into Montrose. Another climb was necessary to Kinnaber Junction, where a speed limit of 15 m.p.h. awaited, and where access was gained to the C.R., later L.M.S., line into Aberdeen. The gradients in this instance were 1 in 100, 1 in 90 and 1 in 88. Three miles from Kinnaber Junction the ascent up the three-mile long Marykirk bank began and the rises of the land here were 1 in 104, 1 in 118, 1 in 98, 1 in 106 and 1 in 139. There was a downward profile past Laurencekirk to Fordoun where the five mile run up to Drumlithie summit began; 1 in 170, 1 in 199, 1 in 228 and 1 in 101 formed the gradients for this part of the line. The track took a downward trajectory to Stonehaven and from there the ground undulates and had a number of sharp curves for a number of miles until a peak is reached past Portlethen, seven miles from Aberdeen. The steepest gradients on that part were 1 in 91, 1 in 93 and 1 in 100. The track then swept down into Aberdeen.

Running from Aberdeen to Edinburgh the line also featured as many severe gradients as travelling in the 'down' direction. The climb out of Aberdeen to Cove Bay is one notable instance with rises of, amongst others, 1 in 96, 1 in 105, 1 in 118 and 1 in 116, while the climb from Stonehaven to Drumlithie is another. The gradients for this latter section were; 1 in 103, 1 in 85, 1 in 92, 1 in 107 and 1 in 102. Between Dundee and Edinburgh, gradients of 1 in 111, 1 in 95 and 1 in 105 took the line from Ladybank Junction to Lochmuir box. Similarly, from Thornton Junction a steep bank had to be tackled after the speed restriction had been observed. The worst gradients on this section were two 1 in 70 stretches split by a brief piece of level land between Inverkeithing and North Queensferry before the Forth Bridge, occurring again after a speed restriction.

To handle the passenger traffic over the route, as well as trains from Edinburgh to Carlisle and Glasgow, the N.B.R. requested a suitable design from William Reid, the company's Locomotive Superintendent, in 1905. The design accepted was for an Atlantic locomotive with two cylinders, 20 in. diameter by 28 in. stroke (the diameter later being increased to 21 in.), Stephenson link motion and 10 in. piston valves (short travel). The boiler was 5 ft 6 in. diameter, initially not superheated, with a Belpaire firebox. Fourteen locomotives, classified 'H' by the N.B.R.,

The Reid Atlantics were the principal express locomotives until the late 1920s. No. 9903 Cock o' the North *was also involved with the feedwater heater experiments and has a Worthington device fitted here.*

were constructed to begin with by the N.B.L.C. at their Hyde Park Works in mid-1906. *Locomotives of the L.N.E.R. Part 3A* (1979) relates that at first there were concerns from the N.B.R. Civil Engineer James Bell about the oscillation of the engines and the effect this was having on the lines that they were running on. H.A. Ivatt of the G.N.R. and V.L. Raven of the N.E.R. both made reports on the running of the class and suggestions as to their improvement, but Raven was particularly impressed by the locomotives. However, none of the suggested minor modifications appear to have been implemented and the concerns dissipated. The N.B.R. subsequently conducted trials with a Highland Railway 'Castle' Class 4-6-0 and London & North Western Railway 'Experiment' Class 4-6-0 in 1910 to see if the wheel arrangement would be more suitable to their lines than the Atlantic, but it would appear that it was not as six more Class H engines were ordered and then built by Robert Stephenson & Co. Ltd in 1911. Two more locomotives were constructed in 1921 by the N.B.L.C. and brought the class total to twenty-two. The final two locomotives were built with superheaters, but the alteration of the earlier saturated engines to feature a 24 element superheater had begun in 1915. Many members of the class had been changed by Grouping and these engines became L.N.E.R. Class C11, while those still to be modified were designated Class C10. The L.N.E.R. enforced this modification and the last C10 had been dealt with by June 1925; all becoming C11.

A H Class Atlantic locomotive was first allocated to Aberdeen in 1907, while the engines working from Edinburgh were initially housed at St Margaret's shed. But, in 1908 the four Atlantics allocated there were transferred to Haymarket. The pattern of working at this time usually dictated that the engine would travel through to its destination before being serviced and returning with another train. From 1919, when working hours were restricted, engines had to be changed at Dundee and three Atlantics were transferred to Dundee at this time for that purpose. The Atlantics were mainly employed on the main expresses, but could also be found working secondary services or fish trains. As mentioned previously, the Atlantics were aided in their sphere of activity after Grouping by the D11/2 Class engines and then the D49 4-4-0s after their introduction. As the 1920s progressed the double heading of the Atlantics with either another member of the class, a D11/2, D49 or other 4-4-0 class was often necessary because of the increase of train weights, coupled with the difficult nature of the line.

This matter was acutely highlighted when the third class sleeping carriages were introduced by the L.N.E.R. in 1928, taking loadings well over what was acceptable for the motive power in operation between the two cities. The result was that Gresley A3 Pacific no. 2573 *Harvester* was tested for clearance over the Edinburgh to Aberdeen line as far as Montrose in 1928 in preparation for the deployment of Pacifics on the route. The locomotive was found to be acceptable over this section, but for the Pacifics to be allowed from Montrose to Aberdeen the L.M.S. were

required to strengthen a number of bridges on that section. This task was not completed until 1930.

A1s, nos 2565 *Merry Hampton*, 2566 *Ladas* and 2567 *Sir Visto*, were subsequently sent from Haymarket to Dundee Tay Bridge shed to work some of the services, while three A3s, new from Doncaster, replaced them at Haymarket. The Pacifics were also limited as to the weight of trains they were authorised to haul between the two cities and these were different for the A1s and A3s. The A1s could take 430 tons tare from Edinburgh to Aberdeen, while for the A3s this was 480 tons tare. In the reverse direction the permissible weights were 370 tons and 420 tons tare respectively. Double heading of the Pacifics over the line was prohibited and the double heading of Atlantics had to be perpetuated when loads exceeded the Pacifics' limits. In comparison with the Pacifics' authorised loads, the C11 Atlantics could take 370 tons tare northwards and 340 tons tare southwards. The D49s could take 340 tons tare to Aberdeen.

A number of runs, by various engines, over the Edinburgh to Aberdeen line have been recorded by writers for a number of publications. Several journeys by the C11 Atlantics were recorded in *The Meccano Magazine* of June 1935 and it is worth recounting some of the salient details here. No. 9869 *Bonnie Dundee* was in charge of the 12.45 p.m. express passenger train, weighing 330 tons tare, 355 tons gross, running on the Aberdeen to Dundee section with driver Campbell and fireman Taylor (both of Dundee shed) at the controls. Starting from the station the regulator was opened three-fifths and cut-off was 53%. After a mile this was reduced to 44% and at Craiginches the speed had risen to 36 m.p.h. Yet, on the ascent to Cove Bay the cut-off had to be increased to 53% and on the climb the speed fell to 30½ m.p.h. From Aberdeen to the top of the bank, the time taken was 14½ minutes. At this point the regulator was closed to one-quarter open and cut-off set at 44% and the locomotive proceeded to Stonehaven, which was reached on time. Three-fifths regulator and 53% cut-off was again used at the start and up to Drumlithie summit, where the speed of the locomotive reached 30 m.p.h. The time taken to complete this section was 11 minutes. For the heavy run out of Montrose, three-quarters regulator with 60% cut-off was used and by Usan the speed had reached 30 m.p.h., with full boiler pressure. The locomotive breasted the summit at 35 m.p.h. and on the descent *Bonnie Dundee* reached 68 m.p.h. past Lunan Bay. A permanent way check at Cauldcoats briefly interrupted progress and then the restriction at St Vigean's Junction was observed. Arbroath, 13¾ miles from Montrose, was reached in 21½ minutes. On the following level section to Dundee three-fifths regulator with 44% cut-off were used and a high of 60 m.p.h. was attained. The section, 17 miles, took 21½ minutes to complete, while the 70¼ miles between Aberdeen and Dundee had taken 105½ minutes of running time.

In the same article a second journey between Dundee and Aberdeen is presented, but this time featuring no. 9509 *Duke of Rothesay* and driver Moodie at the regulator.

He drove the locomotive in a different manner to driver Campbell, as he had the regulator opened full for the majority of the journey and used shorter cut-offs with the result that higher speeds were attained up the banks. Signals also stopped the engine on some sections so a fast running time was required if the locomotive was to keep on schedule. The running time was ninety-five minutes, five minutes ahead of the prescribed time. A brief description of a journey with two locomotives double heading the train is also included. *Duke of Rothesay* was paired with N.B.R. 'Scott' 4-4-0 no. 9340 *Lady of Avenel* for the Dundee to Aberdeen portion of the 'Aberdonian', which weighed 401 tons tare and 420 tons gross. A good performance was garnered from the engines in this instance as departure from Dundee was made four minutes late, but the locomotives succeeded in bringing the train into Aberdeen three minutes before the scheduled arrival.

The N.B.R. Atlantics were reputedly heavy coal consumers and this is borne out by the figure for the year 1937 reproduced in *LNER* (1986), when an average of 68.2 lb was used per mile. However, it must be mentioned that the majority of the class had been withdrawn by the end of 1936 so how accurate a representation of the class as a whole this figure is must be open to debate. Nevertheless, the figure was 10 lb in excess of the average of the C2 Atlantics with the second highest coal consumption and over 14 lb more than the C1 and C7 Classes. It is also interesting to note that some of the other classes at work in Scotland consumed greater amounts of coal than their counterparts employed in England. The D11/2 used just under 12 lb of coal per mile more than the D11/1s working on the G.C. section, while the D49s north of the border burned nearly 14 lb of coal more. Furthermore, the A1 Pacifics were 10 lb heavier on coal. Also in *LNER*, the point is made that the coal used in Scotland was of a lower calorific value than the coal generally available for locomotives in England and this could increase 'the specific weight of coal burned by 10%.' The maintenance required by classes employed in Scotland could also be expected to be higher than those working in England. *Locomotives of the L.N.E.R. Part 3A* (1979) makes such a point in relation to the C11 Atlantics and those formerly of the G.N.R., G.C.R. and N.E.R.: '... they [the C11s] suffered great frame stresses...it was found that engines so employed [on express passenger services on the N.B.R. main lines] required more attention than those of similar type used elsewhere on the L.N.E.R.'

C11 Atlantic no. 9904 Holyrood *rests at the west side of Haymarket shed, Edinburgh. The engine was in service between August 1911 and August 1936.*

Construction

Evidently the introduction of the A1/A3 Class Pacifics on the Edinburgh-Aberdeen line had done little to ease the situation, as by 1931 the Scottish Area Operating Department had appealed to the Locomotive and Traffic Committee for the provision of a suitable locomotive to handle the heaviest trains. In early 1932 the committee met and provision was made for a new locomotive to satisfy the Operating Department's request and the first outline drawing for the new locomotive was produced at King's Cross at the end of March 1932. This resembled the appearance of the A1/A3 Class Pacific engines, but the bogie was replaced by a leading pony truck and an extra pair of coupled wheels, 6 ft 2 in. diameter instead of the 6 ft 8 in. of the Pacifics, was present. A double chimney, three cylinders of 21 in. diameter by 26 in. stroke, and an A3-type 220 psi boiler were the other features of the drawing. Gresley chose the 2-8-2 wheel arrangement because of the additional adhesion this would offer both for travelling on the difficult terrain and for transmitting the required power to the track. Considerations of providing a bogie were dismissed early on because this would have made the locomotive too big for the turntables in use between Edinburgh and Aberdeen. Further, the axle loads would have been altered and reduced the adhesion factor of the coupled wheels. The decision to use the Mikado wheel arrangement was supported by testing one of the P1 freight engines on a passenger service. Gresley had P1 no. 2394 attached to the 7.45 a.m. King's Cross to Doncaster stopping service, the engine in charge of the train as far as Peterborough, and the locomotive handled the train with ease - attaining a top speed of 65 m.p.h. on the journey. The Mikado wheel arrangement had been extensively employed in America and Europe for passenger services, but this was to be the first time it had been employed in Britain to an express passenger locomotive.

Doncaster Drawing Office produced the next drawing in April 1932. At this time a diagram of the proposed

The front end of no. 10000 inspired the design for the new locomotive.

wheelbase, which had been reduced by six inches from the original drawing, and axle loadings, estimated to be 20 tons, was submitted to the District Engineers for the three sections of the L.N.E.R. for approval. The proposal would have also been given to the relevant department in the L.M.S. as the engine would be travelling over the company's line into Aberdeen. Subsequently, two engines were authorised to be built to these basic specifications in

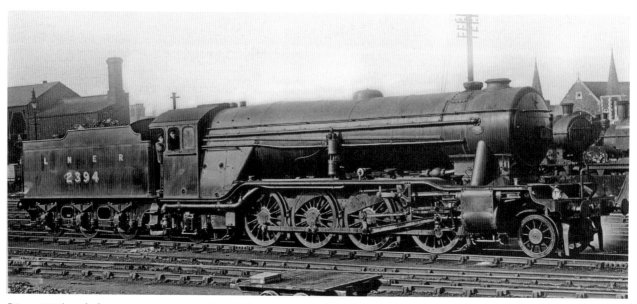

P1 no. 2394 worked a passenger train to test the wheel arrangement and was found more than satisfactory.

The locomotive's frame plates in the slotting machine at Doncaster Works.

The frames receive the attention of a worker and his cutting torch.

February 1933. However, this was revised to one engine and order no. 330 was issued to Doncaster Works in March 1933. A further design permutation occurred at this time. The length of the engine was enlarged by 1 ft 3¼ in. because of an extension in the size of the firebox and this also meant that the weight of the locomotive would be increased by 3 tons 9 cwt. By mid-1933 the new engine had evolved substantially more to include the Kylchap type of double blastpipe and chimney, with smoke deflectors necessarily being employed. These were of a similar design

to those fitted to W1 4-6-4 no. 10000, which had been developed by Professor W.E. Dalby of the City & Guilds College using a scale model of that locomotive in a wind tunnel, and had been proved to be satisfactory in service. Also included in the plans were an A.C.F.I. feedwater heater, perforated steam collector and Lentz poppet valves with rotary cam valve gear. The new diagram, produced to show all of these additions, illustrated that the total length of the locomotive had increased to 73 ft 8⅜ in. and the estimated weight of just the engine was 110 tons. The final

The frames have been assembled over a vacant pit in the New Erecting Shop, Doncaster Works, February 1934.

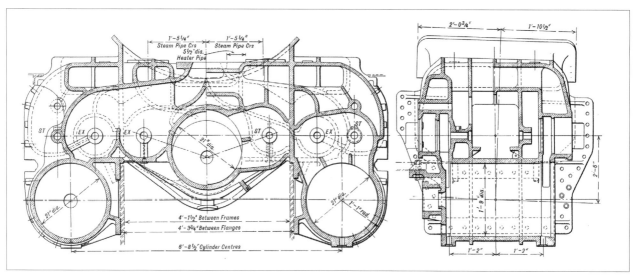

The centre front and side section drawings of the monobloc cylinder casting.

addition to the design was the v-shaped cab front. This had been developed by the P.L.M. in France as a method of reducing wind resistance and also affording a better view forward, especially at night, for the footplatemen.

Construction of the new locomotive began in late 1933. For the main frames, two 36 ft 9½ in. sections were cut out of 1⅛ in. steel plate and these were then placed in Doncaster Works' erecting shop for assembly. Two additional frame sections were also produced and added to the front and rear of the main frames. These were 13 ft 11¼ in. and 11 ft 3¼ in. long respectively; for the rear frame section the thickness was ⅛ in. less than the two other sections. The front addition to the main frames was connected between the leading and second pair of coupled wheels, while the rear frame section was attached near to the position of the firebox. The spacing of the right- and left-hand sections of the frames differed over their 45 ft 5¼ in. length. The front

section was 4 ft 1½ in., the main frames 3 ft 11¼ in. at the top and 3 ft 5 in at the bottom and the rear section was 6 ft 0½ in. apart. Where the main frames tapered inwards a 1 in. thick steel plate was fixed to the frames on the outside to add strength to this section. Cast-steel was utilised for a number supplementary pieces attached to the frames, such as; the front pony truck stretcher and other frame stretchers, the dragbox, running plate support brackets, buffer brackets, the inside motion plate and smokebox saddle and outside motion plates. In relation to this latter component, the positioning was dictated by the placement of the leading coupled wheels and the motion plates had to be attached further back than would have been desirable. As a result the casting was of a complicated pattern and it projected out past the first two pairs of coupled wheels. The axlebox hornblocks were also fixed to the frames at this point and their alignment was carefully measured.

A view of the front of the casting, February 1934.

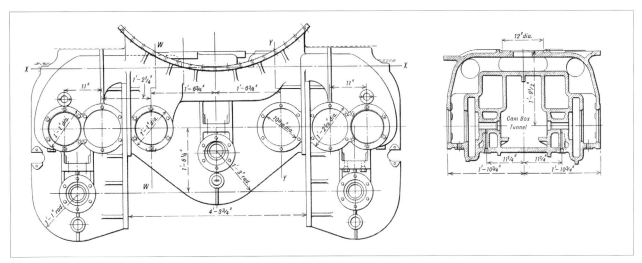

The rear of the cylinder casting with a section drawing of the steam valve for the centre cylinder from point W. to W.

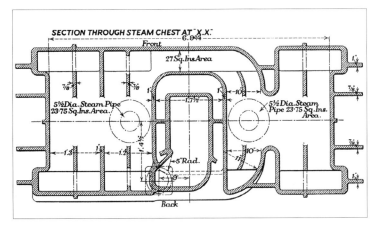

Above: *A longitudinal section drawing of the steam chest at the top of the casting from point X. to X.*

Right: *The side of the cylinder casting with the tunnel for the cam box.*

Looking at the rear of the cylinder casting.

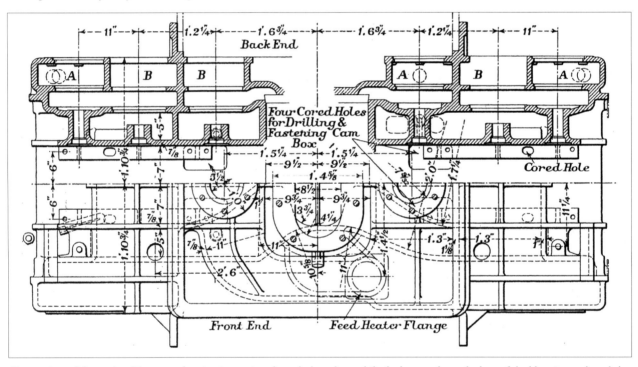

Two sections of the casting. The upper drawing is a section through the valves, while the bottom shows the base of the blastpipe and smokebox saddle flange.

The cylinders and valves for the locomotive were cast together to form a 'monobloc' cylinder unit. However, this was not produced by Doncaster Works, but at Gorton Works, Manchester, as the works provided the L.N.E.R's cylinder castings. The production of the monobloc casting was partially the responsibility of R.A. Thom, Mechanical Engineer, Doncaster, and the piece was probably one of the most complicated casting of its kind to be produced at the time because of the greater number of valves used in the poppet valve system. Twelve valves were provided in this instance, with four serving each cylinder - two for

steam inlet and two for exhaust - and space also had to be provided for the passageways and ports. The monobloc unit weighed approximately 7 tons. The steam valves were 8 in. diameter and the exhaust valves 9 in. with large ports provided to allow free steam flow. The outside cylinders were centred at 6 ft 8½ in., while the width of the unit was 8 ft 11¾ in. The outside cylinders were inclined at 1 in 30, but the inside cylinder was inclined at 1 in 7.713 in order for it to clear the leading coupled axle and for it to drive on to the second coupled axle as per Gresley's preference. The monobloc casting was attached to the

The cylinder casting attached to the frames. Note also the cast-steel outside motion plate and buffer brackets.

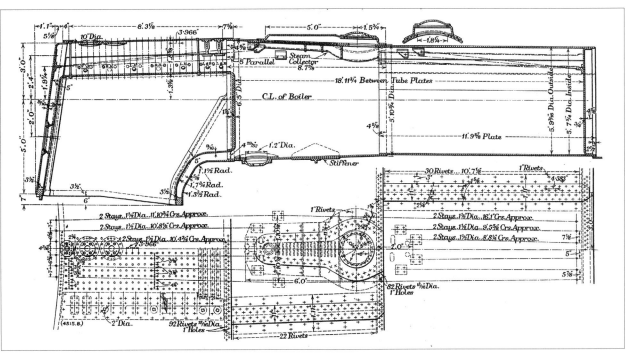

The diagram 106 boiler. The top drawing is a section through the boiler. The bottom shows the top of the two sections of the boiler barrel and the arrangement of the firebox stays.

frames by flanges 1⅜ in. thick and 4 ft 3¾ in. apart. The cylinder walls were 1½ in. thick, steam inlet valve walls were 1 in. deep and the exhaust walls were ⅞ in. thick.

The next step in the erection of the engine was the introduction of the boiler to the frames and cylinders. The boiler used, diagram 106, was very similar to the diagram 94HP in use on the A3 Pacifics (and the 94A which was about to be introduced on the final nine new locomotives of the class). At the firebox end the boiler's diameter was 6 ft 5 in. and at the smokebox end this had narrowed to 5 ft 9³⁄₁₆ in. inside diameter. The boiler barrel was made

of two sections - a taper and a parallel portion (the metal plates being ²³⁄₃₂ in. and ²⁵⁄₃₂ in. thick) - and was 19 ft. total length, containing 121 tubes, 2¼ in. diameter and 10 S.W.G. As built, the boiler was fitted with 'sine wave' superheater elements, which rippled along their length to move the steam through the element and increase the transfer of heat to the steam. There were 43 superheater flues, 5¼ in. diameter and ⁵⁄₃₂ in. thick, with 43 'sine wave' superheater elements, 1¼ in. outside diameter and 11 S.W.G. The distance between the tube plates was 18 ft. 11¾ in. and the firebox tubeplate protruded into the

Boiler no. 8771, pictured in February 1934.

The smokebox end of boiler no. 8771.

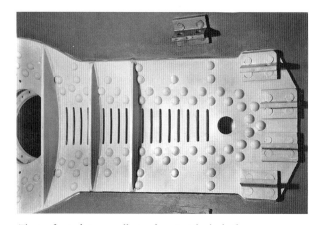

The perforated steam collector from inside the boiler.

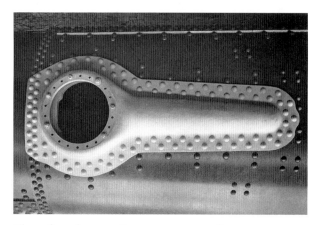

The perforated steam collector as seen externally.

The top of the boiler as viewed from the cab end. Note the mixing and settling chambers of the A.C.F.I. feedwater heater at the front of the boiler.

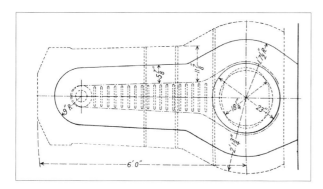

barrel to create a combustion chamber 3 ft in length. The firebox itself was 10 ft 9 in. long outside and 7 ft 9 in. width at the bottom, while the inside was 9 ft 2¾ in. long at the top. The firebox crown was 6 ft 8¹³⁄₁₆ in. above the foundation ring at the front and at the rear was 6 ft 0⁵⁄₁₆ in. The copper plates used in the construction of the firebox were ⁹⁄₁₆ in. thick. The firebox grate was 7 ft 2 in. long and 6 ft 11¾ in. wide, with these combining to give a grate area of 50 sq. ft. The heating surface of the firebox was 237 sq. ft, the superheater flues provided 1,122.8 sq.

Boiler no. 8771 in the New Erecting Shop in March 1934 with the smokebox.

The boiler and cylinders were lagged with 'Alfol' insulating material.

ft and the 'sine wave' superheater elements 695 sq. ft.; the total heating surface of the boiler was 3,409 sq. ft. Two 3½ in. diameter Ross 'pop' safety valves were used and the working pressure was 220 lb per sq. in.

The boiler, no. 8771, was also noteworthy because it was fitted with a perforated steam collector behind the steam dome; the shape of the whole arrangement giving rise to the colloquial 'banjo dome' moniker. The perforated steam

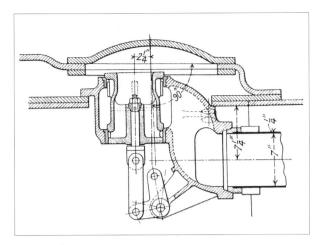

The regulator.

collector was employed to remove any water remaining in the steam before entering the regulator. This was done by passing the steam through 17 half-inch wide slots cut in the top of the boiler barrel. The barrel was reinforced at this point by a large metal plate with corresponding holes and three support ribs of ¾ in. plate. The exterior casing was made from a steel pressing, which was integral with the steam dome and was riveted to the boiler barrel from the exterior; the length of the perforated steam collector was approximately 6 ft. Gresley had experimented with a steam dryer a number of years prior to this application and the former had been quickly discarded. He evidently

became satisfied with the perforated steam collector as it was applied by him to the boilers of the final nine new A3s, the A4 Class, and V2s.

The regulator was of the balanced type, then being introduced by Gresley, and the main steam pipe was 7 in. diameter. Two pipes of 5½ in. diameter took the superheated steam to a large steam receiver positioned on top of the cylinder casting and extended over the outside cylinder centre lines. Particular attention was paid to the finish of the steam passages from the regulator to the cylinders. The metal inside the pipes was made as smooth as possible to aid the free flow of steam and to reduce instances of drops in pressure while the steam was on its way to the cylinders. As mentioned previously, this was one of Chapelon's principles employed in his rebuilds and the idea was enthusiastically embraced by Gresley. In his Presidential Address to the Institution of Locomotive Engineers, made in September 1934, Gresley comments upon this connection. 'The engines of the Paris Orleans Railway...have achieved results in the haulage of long-distance high-speed trains of great weight over a severely-graded line which had never been attained by engines of similar weight. In preparing the designs of the new eight-wheel coupled express passenger engine recently constructed at Doncaster, I did not hesitate to incorporate some of the outstanding features of the Paris Orleans engine, such as the provision of extra large steam passages and a double blast-pipe.'

A view of the locomotive from the front showing the lagging around the cylinder.

The locomotive's smokebox, showing the Kylchap cowls and chimney. Also note the Crosby tri-note whistle positioned in front of the chimney.

The locomotive was the first on the L.N.E.R. to carry the Kylchap double blastpipe and chimney, although a D49/1, no. 251, and D49/2, no. 322, had been fitted with a single chimney version in 1929 (Chapelon had been on the footplate of no. 251 to comment on the application), but the apparatus was later removed from both. As applied to the new locomotive the exhaust arrangement consisted of two exhaust pipes, 6½ in. diameter, leading from an exhaust chamber, similar to the steam receiver, to the

blastpipe tops. To obtain an adequate spread of steam over the orifice area the pipes had ⅜ in. diaphragms and there were also curved ribs of ⅞ in. thickness in the exhaust passages to aid the flow of the exhaust. The blastpipe tops were removable to allow the most suitable dimension for the locomotive to be fitted. The diameter of the blastpipe orifice could be altered, as could the size of the taper blocks, also referred to as vee bars, present in the blastpipe top, which split the exhaust as it exited the orifice. As built

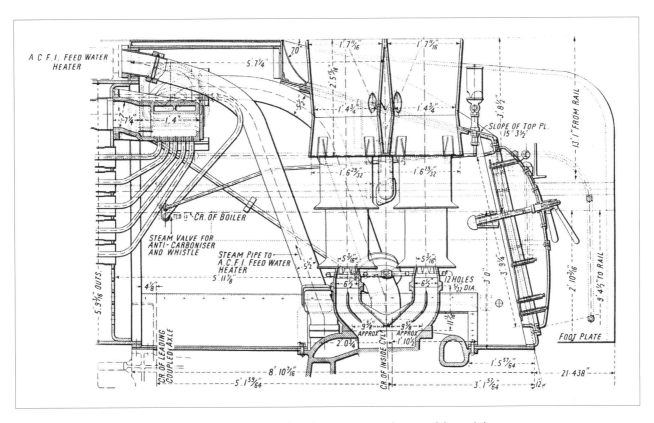

Above: The arrangement of the smokebox as seen from the side. Below: A cut-away drawing of the smokebox.

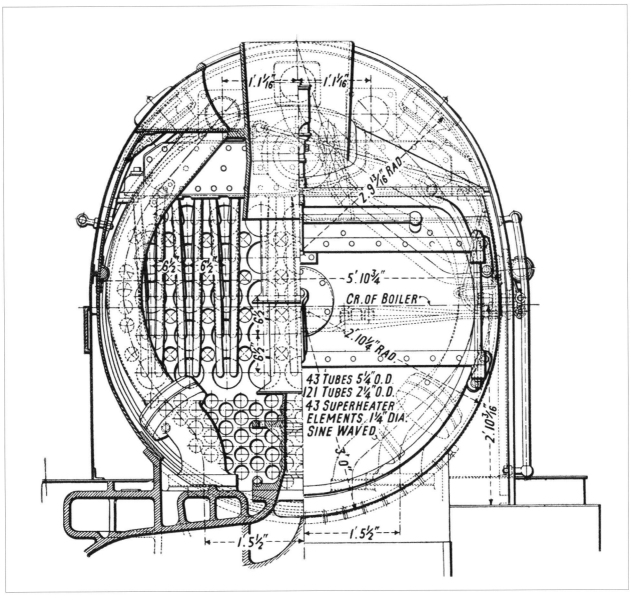

The engine is lifted for the coupled wheels to be fitted in April 1934.

the engine had two blastpipe orifices of 5³⁄₁₆ in. diameter with no. 3 taper blocks fitted, which was the largest size available, and the area through each blastpipe orifice was 16.4 sq. in. From the blastpipe orifices the exhaust would pass into two Kylälä spreaders, which were flared at the bottom to a diameter of 1 ft 3¾ in. and then closed into four convergent tubes, with the assembly being 1 ft 3⁹⁄₁₆ in. tall. The spreaders then discharged into two cowls fixed above them that were 1 ft 1³⁄₈ in. diameter, 1 ft 1³⁄₈ in. high. The chimney cowls converged before diverging and were 1 ft 6²⁹⁄₃₂ in. diameter at the bottom, 1 ft 4¾ in. at the choke and 1 ft 7¹¹⁄₁₆ in. diameter at the top. The chimney cowls were 2 ft 5¹⁵⁄₁₆ in. tall and the distance from the bottom of the cowls to the choke was 10¼ in. The centre line of the chimneys was 5 ft 7¼ in. from the rear of the smokebox. The smokebox door and front plate were inclined at 12 and 15 degrees respectively, as opposed to the 18-degree slope employed by no. 10000, to throw the exhaust clear of the locomotive.

The whistle had to be moved from the traditional position at the cab end and was resituated in front of the chimney. The whistle was a new type to that normally used by L.N.E.R. engines and had been given to Gresley by Captain Howey of the Romney, Hythe and Dymchurch Railway. Two Crosby tri-note whistles had been acquired for use on two of the railway's new locomotives - no. 9 *Dr Syn* and no. 10 *Black Prince* - as they had been modelled on Canadian Pacific Railway locomotives which used the Crosby type. Spencer (1947) notes that Gresley was eager for his new locomotive to 'have a distinctive crow' and mentioned this to Captain Howey, who presented him with one of the whistles.

An A.C.F.I. feedwater heater was also fitted to the locomotive. The tandem pump was affixed to the running plate on the right-hand side. The mixing chamber and second chamber were attached to the parallel section of the boiler behind the smokebox and were obscured from view by the boiler clothing plates. Exhaust steam, from both the cylinders and pump, was delivered to the mixing chamber, with the former arriving through a 5½ in. diameter pipe. The mixing of the steam and feedwater was performed in the same manner as on the A1/A3 Class engines. The hot water was then fed into the boiler on the right-hand side. A Gresham & Craven no. 11 live steam injector was also fitted to the engine.

The following stage in the locomotive's construction was for the pony truck, coupled wheels and trailing wheels to be fitted. The pony truck was of Gresley's double-bolster swing link type and had 3 ft 2 in. diameter wheels. The axle journals were 6½ in. diameter by 9½ in. long. The wheels were confined to 4½ in. movement forwards or backwards of their central position. Helical springs were used, these being 10³⁄₁₆ in. long, when not compressed (the vertical deflection of both springs was 0.17 in. per ton), 5½ in. outside diameter and of Timmis section. The distance between the swing link pin centres was 7 in.

The coupled wheels were 6 ft 2 in. diameter with 9½ in. by 11 in. axle journals and laminated springs were used, these consisting of 15 plates 5 in. wide by ½ in. thick, positioned at 3 ft 6 in. centres. The tyres on the middle pair of coupled wheels were slightly thinner than the other wheels' tyres as an aid to helping the engine negotiate the many curves on the Edinburgh to Aberdeen line. The rear axle had 3 ft 8 in. diameter wheels, Cartazzi

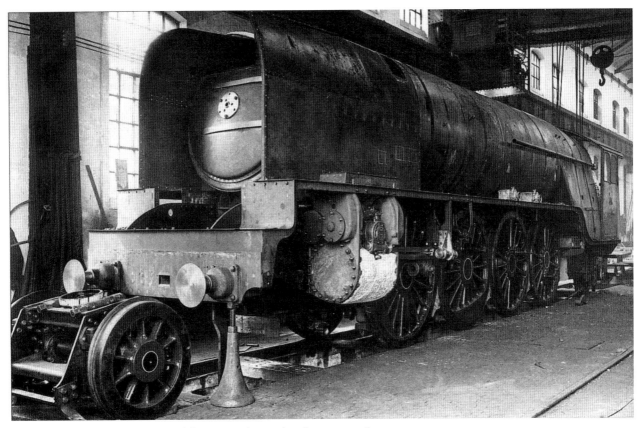

The coupled wheels are in position and the pony truck is ready to be put in its place.

axleboxes and 6 in. by 11 in. journals. Laminated springs were again used with 14 plates, ⅝ in. thick and 5 in. wide, centred at 4 ft 6 in.

The spacing between the pony truck wheels and the first pair of coupled wheels was 8 ft 11 in., 6 ft 6 in. for all the coupled wheels and 9 ft 6 in. between the rear coupled wheels and the trailing axle. The coupled wheelbase was 19 ft 6 in, while the total engine wheelbase was 32 ft 11 in. The leading axle supported 11 tons 10 cwt, the coupled wheels, respectively, 19 tons 12 cwt, 20 tons 10 cwt, 20 tons 10 cwt, 20 tons and the rear axle 18 tons 3 cwt. The locomotive's adhesive weight was 80 tons 12 cwt and had an adhesion factor of 4.15. The tractive effort of the engine was 43,462 lb. Crank pins for the outside cylinders were 6¾ in. diameter by 6 in. long, while the inside crank pin was 9¼ in. by 6 in. For the coupled wheels the coupling pins were, respectively, 4¾ in. by 4 in., 7½ in. by 4½ in., 4¾ in. by 5 in. and 4¾ in. by 5 in. Coupling and connecting rods were manufactured from nickel chrome steel. The third pair of coupled wheels drove a return crank (featuring roller bearings on its linkages) which in turn operated the two Wakefield mechanical lubricators used to supply oil to the cylinders and axleboxes.

The valve gear employed by Gresley to operate the double beat Lentz poppet valves was the rotary cam arrangement similar to that used by the D49/2 Class engines. However, due to the placement of the three cylinders two camshafts had to be used and they operated six valves each. The camshaft working from the right-hand side operated the four valves for the outside cylinder and the two steam inlet valves for the inside cylinder. The left-hand side camshaft operated the four valves for the left-hand side outside

cylinder and the exhaust valves for the centre cylinder. The rotary motion for the camshafts was taken from a return crank on the main crank pin which turned bevel gears connected to a driving shaft that imparted the movement to further gears coupled to the camshaft. The camshafts, housed in cam boxes, were placed in a hollow, present in both sides of the cylinder casting, and they could be easily removed through a door if needed. The camshafts featured six scroll cams each for the admission and exhaust of steam from the cylinders. The cams allowed an infinitely variable number of cut-off positions between 10% and 70% and 31% and 70% in reverse gear. Reverse gear was effected by a wheel in the cab which was connected to a number of gears and shafts, with the latter being visible on the left-hand side of the boiler, and the camshafts were moved to bring different areas of the cams into use. As in the other applications of Lentz poppet valves, the scroll cams allowed the steam valves to be closed and the exhaust valves to be open when the motion was placed in mid-gear. The cylinder clearance volumes (the volume of space between the cylinder and piston when the piston is at top dead centre, including the port area) were; for the outside cylinders 12.4% at the front and 11.78% at the rear. For the inside cylinder the clearance volume was 16.1% at the front and 15.8% at the back.

The tender was another departure from L.N.E.R. practice as it was fabricated using welding. In *Bulleid of the Southern* (1977) the point is related that Bulleid was a strong proponent of this method from the early 1930s as it reduced costs and saved weight by comparison with riveting, which was the main method of construction at the time. At the time of the tender's erection, the L.N.E.R.

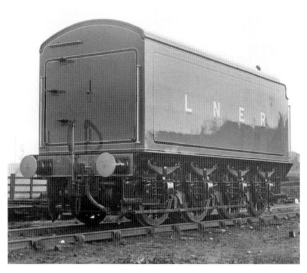

New type non-corridor tender no. 5565 was the first to be built for the L.N.E.R. using welding. These two pictures were taken in April 1934 at Doncaster Works.

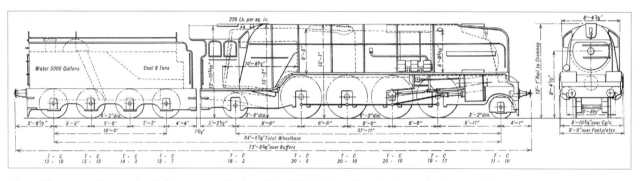

Above: The engine diagram for the locomotive as built in May 1934. Below: Completed outside the Paint Shop.

An unidentified group have their picture taken with Cock o' the North. *The men on the extreme right and left are the two chargehand fitters seen in the picture on page 72.*

was testing welding on wagons, carriage underframes and an order had been placed with Babcock & Wilcox for an all-welded boiler; this later being fitted to a J39 0-6-0. The new locomotive's new type non-corridor tender, no. 5565, had been ordered from Doncaster early in 1933, but by the mid-part of the year Metropolitan Vickers had been approached and commissioned to fabricate the tender's body. The base plate was 24 ft 4¼ in. by 8 ft 9 in. and the side and rear sheets were 7 ft 0⅛ in. tall. The coal space was welded into position to form the tank. The division plate arched from one side to the other and at its centre the height was 7 ft 10⅜ in. above the level of the footplate. The tender tank had a 5,000 gallon water capacity and the coal space could accommodate a nominal 8 tons. Doncaster Works married the tender body to the frames, which were of 1 in plate and 24 ft 1½ in. long. The wheels were 4 ft 2 in. diameter and spaced (front to back); 5 ft 3 in. 5 ft 6 in. 5 ft 3 in. and 5 ft 6½ in., with the total wheelbase being 16 ft. The total weight of the tender, when at its water and coal capacity, was 55 tons 6 cwt. The axle loads, when full, were; 13 tons 7 cwt, 14 tons 3 cwt, 13 tons 18 cwt, and 13 tons 18 cwt. The tender was 2 tons 12 cwt lighter than its counterpart manufactured through riveting. Another notable difference of the tender was that the wheels used had spoked centres instead of disc centres.

The final stage of the engine's construction was for the L.N.E.R's apple green livery with black and white lining to be applied. The boiler had been lagged with 'Alfol', which was an insulating material made from aluminium, and this extended to between the smokebox and smoke deflector plates, allowing the latter to have the apple green livery applied. The locomotive was numbered 2001 after the original number, 2981, was discarded. The former had been carried by a N.E.R. Class S 4-6-0, L.N.E.R. B13, which incidentally had been the first 4-6-0 to run in Britain and had been withdrawn in June 1931. C.J. Allen, in his article on the locomotive which appeared in *The Railway Magazine* of July 1934, writes that Gresley had informed him that the use of the number was 'more by accident that design.' The name given to no. 2001, *Cock o' the North*, was used more deliberately because of the name's strong connection to the engine's area of operation. However, the name was already in use at this time by N.B.R. Atlantic no. 9903 and had to be stripped from the locomotive at the beginning of May 1934; the engine received *Aberdonian* as a replacement.

No. 2001 *Cock o' the North* was completed at Doncaster Works on 22nd May 1934 and classified P2/1.

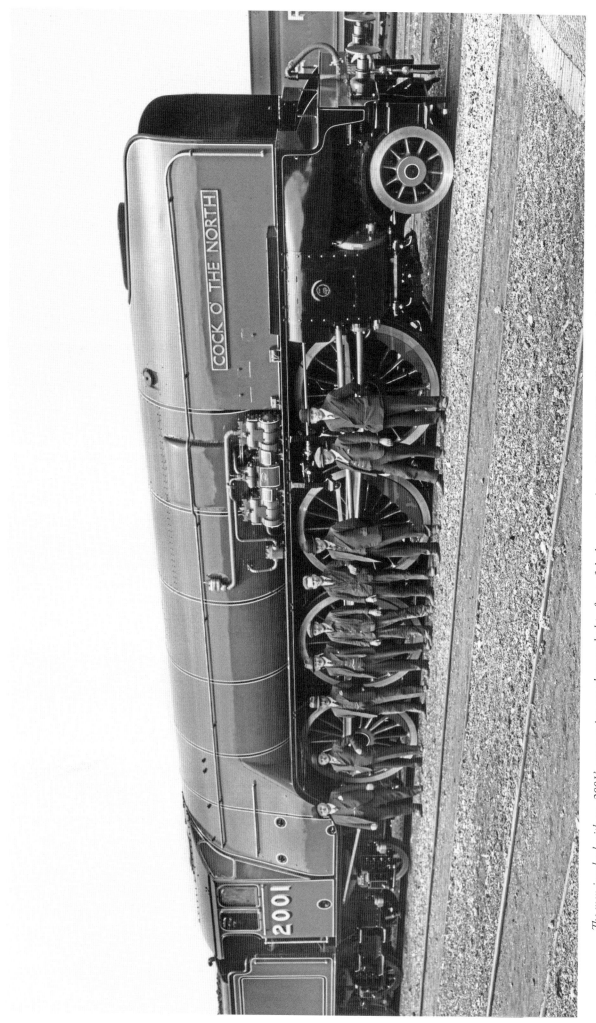

The men involved with no. 2001's construction are photographed in front of the locomotive during May 1934. The officials, from the left, are: R.A. Thom, Mechanical Engineer, Doncaster, E. Windle, Chief Locomotive Draughtsman, J.S. Jones, Assistant Works Manager, F.H. Eggleshaw, Works Manager. The other members of staff, from the right, are; Paint Shop Foreman, unidentified, Erecting Shop Foreman, chargehand fitter and chargehand fitter.

A three-quarters broadside view of Cock o' the North taken in May 1934. Note the reversing shaft above the straight running plate, which also clears the coupled wheels. The nameplates were consequently a departure from standard practice for Gresley's named engines and they became straight and were then mounted prominently at the front of the engine. The linkage to drive the mechanical lubricators from the third pair of coupled wheels is also visible.

The oiling points are checked after leaving Doncaster Works' Paint Shop.

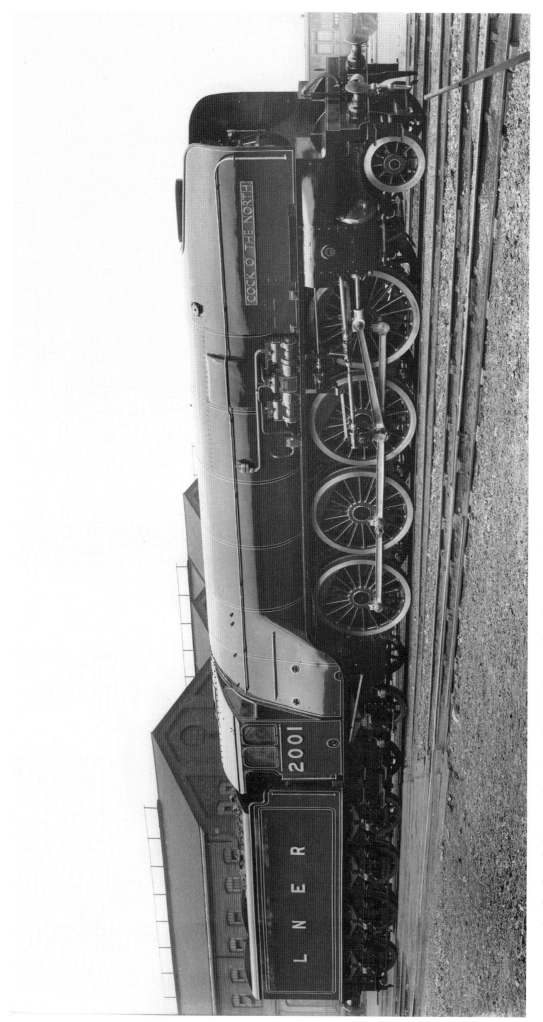

Another broadside view of no. 2001 Cock o' the North. The A.C.F.I. feedwater heater was painted in the same livery as the locomotive. A number of detail modifications were carried out subsequently; a Teloc speed recorder would be fitted to the engine and this was connected to the rear crank pin on this side. Further, the cab side sheets would be reduced to bring them into conformity with the tender contour as a result of the addition of back rests to the enginemen's seats. This would also cause an increase in the length of the cab's vertical handrail.

The cab of no. 2001. The hand-wheel on the left-hand side controlled the cut-off positions through the reversing shaft. A heat shield for the fireman was later provided.

One of the first tasks performed by *Cock o' the North* was to aid the 7,000 railway workers and the townspeople of Doncaster. From Saturday 26th to Sunday 27th May 1934 the locomotive was the star attraction of Doncaster Works exhibition, which was being held to benefit Doncaster Infirmary and railwayman's charities. The *Doncaster Chronicle* of 31st May 1934 states that *Cock o' the North* was joined by a number of other interesting items of L.N.E.R. rolling stock, including; a Sentinel steam railcar, the Beyer Garratt U1 Class 2-8-0+0-8-2, A1 Pacific no. 4472 *Flying Scotsman*, an instruction van containing models of valve gears and a similar one for signalling, a post office van, an invalid or private family saloon, buffet car, Pullman carriages, first and third class sleeping carriages, restaurant carriages and a number of goods wagons. A breakdown crane was also requisitioned to swing a 12-ton wagon around which was filled with people, who had paid two pence for this enjoyment. There were about 40 exhibits in total and they were displayed in the Crimpsall Repair Shop sidings. The event had been organised by a committee of railway officials, with Mr F.J. Trotter, District Superintendent of Doncaster, acting as Chairman. The employees of the works also volunteered to be guides for the spectators or to explain the workings of a number of the exhibits. A brief

opening ceremony occurred on Saturday and this featured Mr Ronald Matthews (later Sir) of the L.N.E.R. Board, the Mayor of Doncaster Councillor Raynard and Lord Lonsdale. Mr Matthews commented that the L.N.E.R's Board were 'very proud' of the work Doncaster carried out for the railway and he paid tribute to Mr Thom and all the staff associated with the works for their service to the company. Mr Matthews also said that the exhibition was a 'welcome opportunity' to show the general public that the railway was 'still able to more than hold their own in regard to moving the population of the country from one part to another.' He added that he was sure the public would agree with his belief that the L.N.E.R. 'headed the field' as a transporter of freight and passengers. After Mr Matthews had also praised the fine work done by Doncaster Infirmary, Mayor Raynard declared the exhibition open. This distinguished party then headed straight for *Cock o' the North* where R.A. Thom explained the 'intricacies of the monster' to Lord Lonsdale, who later 'leaned on the rail of the cab with his head out of the window like an experienced driver.' The attendance for the two-day exhibition totalled almost 40,000 and nearly £1,000 had been raised.

No. 2001 was present at King's Cross station on 1st June for its press demonstration and on the 2nd June the

No. 2001 was the main attraction at the Doncaster Works exhibition in May 1934 and the engine also attended a number of similar events during its career.

Cock o' the North *seen again at the Doncaster Works exhibition.*

No. 2001 in the locomotive yard, on the west side of King's Cross station, in June 1934.

locomotive was at Ilford for a similar exhibition to that described at Doncaster; no. 4472 *Flying Scotsman* was also present. From Ilford, *Cock o' the North* made its way up to Scotland for its first journey between Edinburgh and Aberdeen, which took place on 4th June. The train consisted of 12 carriages and left the capital city at 2.20 p.m. The *Dundee Courier* notes that on its way to Aberdeen *Cock o' the North* spent several minutes at Dundee Tay Bridge

Cock o' the North *at King's Cross station prior to departure from platform eight with a demonstration service for the press on 1st June 1934.*

The young and old admire Doncaster's latest product at King's Cross, 1st June 1934.

station. Then, the locomotive was displayed for the public at Aberdeen Joint station for several hours on 5th June and early in the morning no. 2001 was inspected by the Lord Provost of Aberdeen Sir Henry Alexander with Gresley conducting him. At 2.15 p.m. *Cock o' the North* returned to Edinburgh as it was scheduled to be present at Waverley station on the 6th. The city's Lord Provost, W.J. Thornson, was also introduced to the locomotive during the day.

Following this engagement no. 2001 returned to England to begin 'running in' between King's Cross station and Peterborough. *Locomotives of the L.N.E.R. Part 6B* (1991) records that this comprised of the 10.58 a.m. stopping train or 'Parliamentary train' (which was a cheaper service that was required to stop at all the stations) and the 2.48 p.m. return express. *Cock o' the North* worked this schedule for one week, beginning Monday 11th June and finishing on Saturday 16th June. Driver Charlie Peachey was in charge of the locomotive for this period and he had started his career at King's Cross as a cleaner in 1895, working his way into the top link. In *A Gresley Anthology* (1994) he recollects that the duties were slightly different from the ones quoted above: 'I had *Cock o' the North* first on my normal turns of duty that week, which were a 'Parley' train to Peterborough and back on an express three days, and on the 1.20 as far as Grantham and back with another express on three days.'

On Tuesday 19th June *Cock o' the North* underwent a trial with the dynamometer car in order to test its capabilities. This took place from King's Cross station to Barkston, Lincolnshire and returning to London. The train was made up of a further 18 carriages and brought the total load behind no. 2001 to 649 tons; as in the previous week, driver Peachey was in charge of the locomotive. At 9.50 a.m. this heavy load was started easily by *Cock o' the North* and the drawbar pull at this moment was 14 tons. On the gradient of 1 in 107 to Holloway Junction, through Copenhagen and Gasworks tunnels, the cut-off was 45% and the speed reached 20 m.p.h. Cut-off was then taken to 20% at the top of the rise and Finsbury Park was passed in 6 minutes 6 seconds at 32½ m.p.h. At Hornsey an 8-mile section of 1 in 200 began to Potters Bar and here the speed of *Cock o' the North* was 58 m.p.h. with a 5-ton pull on the drawbar - this equating to 1,730 drawbar horsepower. Halfway up this ascent the cut-off was altered to 22% and the boiler pressure was 190 psi. As no. 2001 breasted the summit the speed was 50½ m.p.h. and the time taken from King's Cross to Potters Bar had been 17 minutes 40 seconds. Hatfield was passed 5 minutes 50 seconds later at 70 m.p.h., 3½ minutes under the allotted time for the whole section from King's Cross.

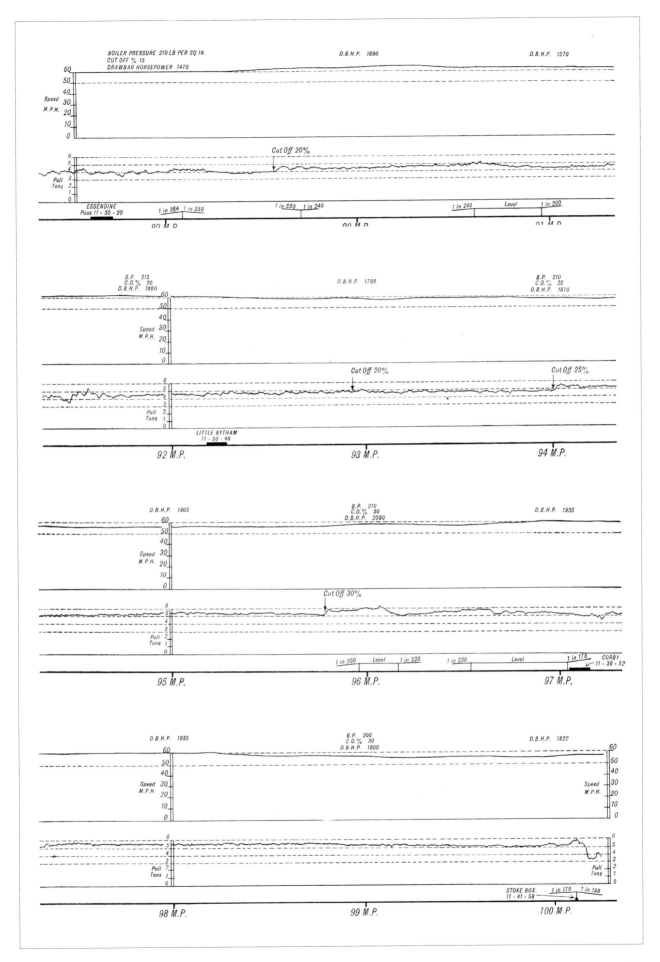

Above: The dynamometer car record for the 11 miles between Essendine and Stoke summit. Cock o' the North *produced 2,090 drawbar horsepower, which was a high figure by British standards at the time.*

Opposite page: On the turntable in the locomotive yard, King's Cross station, June 1934.

Cock o' the North *at the head of an express service.*

Cock o' the North was held back on the following sector from Hatfield to Hitchin, approximately 14½ miles, with the result being that this was completed in only half-a-minute under the 14 minutes allowed - total time 36 minutes 4 seconds - and the greatest drawbar horsepower recorded on this section was 1,300. The next part of the journey was to Huntingdon, which featured 26 miles of track with light gradients allowing the cut-off to be 10%. On the level section of track between Sandy and Tempsford 1,050 drawbar horsepower was produced at 70 m.p.h. After passing Huntingdon station (in 60 minutes 44 seconds from King's Cross), a rising gradient of 1 in 200 awaited *Cock o' the North* and it tackled this at 15% cut-off, reaching the top of the bank, near Abbotts Ripton, at 50 m.p.h. A pull on the drawbar of 4.2 tons or 1,330 drawbar horsepower had been recorded in the dynamometer car. On the run into Peterborough 10% cut-off was used, apart from a

No. 2001 is pictured at Peterborough while being 'run-in' during June 1934.

Cock o' the North arrives at King's Cross in May 1935 with a passenger service from Doncaster. The locomotive worked from the town for a few months upon its return from France before being dispatched to Scotland in June 1935.

period at 30% for a permanent way check at Yaxley and *Cock o' the North* arrived at Peterborough 81 minutes 46 seconds after the start at King's Cross; this would have been 2 minutes quicker if not for the permanent way restriction. The locomotive was in Peterborough station for 2 minutes 25 seconds and this allowed Gresley time to take up a position on the footplate for the remainder of the journey to Barkston.

Upon starting from the station *Cock o' the North* produced a drawbar pull of 16½ tons and at 10 m.p.h. this was recorded as 12 tons. The cut-off position was then progressively shortened to reach 20% at 33 m.p.h. On the 2-mile rise to Essendine, where the gradient was 1 in 264, the speed up the bank was 60 m.p.h., with a 4.1-ton pull on the drawbar corresponding to a drawbar horsepower of 1,470 and the time to the top from Peterborough, 12½

The locomotive works an early evening express to Leeds out of Hadley Wood North tunnel.

No. 2001 has been charged with taking an 'up' express from Edinburgh on the final leg of its journey to King's Cross in mid-1934.

miles, was 15 minutes 5 seconds. For 5 miles beyond this point, the incline of the line increased even further to surmount Stoke summit, which was the highest point of the line between King's Cross and York, reaching 345 ft above sea level. The gradients on this section were 1 in 200, level, 1 in 200, level and 1 in 178. On the first 1 in 200 near Little Bytham the cut-off was 20% and the speed 60½ m.p.h. with a drawbar pull of 4.6 tons and 1,660 drawbar horsepower being developed. At 58 m.p.h. the figures were, respectively; 4.9 tons, 1,700. The cut-off was then increased to 25% and at 58 m.p.h. 5.4 tons was the pull on the drawbar and 1,870 was the drawbar horsepower. The speed then dropped slightly to 56 m.p.h. and the figures then became; 5.4 tons, 1,805 drawbar horsepower. Up to this point the boiler pressure had remained more or less static at 210-215 psi. About a mile away from Corby the cut-off was changed to 30% and on the first level stretch the speed was 57½ m.p.h. with a drawbar pull of 6.1 tons and this produced a drawbar horsepower of 2,090. Corby was passed in 23 minutes 37 seconds from the re-start. On the final section to the summit the speed at the bottom was 60½ m.p.h., while the drawbar pull remained relatively steady around 5.5 tons for the duration of the ascent. The highest drawbar horsepower was 1,935 at the bottom of the climb, while at the top the speed was 56½ m.p.h., the drawbar pull 5.7 tons and 1,920 drawbar horsepower. The 3 miles at 1 in 178 between Corby and Stoke summit had been travelled in 3 minutes 6 seconds. After these exertions *Cock o' the North* made its way steadily into Grantham and from Peterborough, 29¼ miles, the time taken by no. 2001 was 31 minutes 57 seconds, 4 minutes 3 seconds in front of the time allowed.

The train was turned on the triangle of lines 4 miles away from Grantham at Barkston and when starting the pull required was 14½ tons. After a minute had elapsed

the cut-off was reduced to 40% and a minute later the speed was 26 m.p.h., drawbar pull 8.3 tons and drawbar horsepower 1,295. *Cock o' the North* was then run at 15% cut-off passing Grantham at 49 m.p.h., before the cut-off was increased to 20% for the 1 in 198 to Stoke summit. However, similar figures to those going up the opposite side were not achieved and the speed at the bottom was 49 m.p.h., falling to 45 m.p.h. at the top. The highest drawbar horsepower was 1,490, the lowest on this section being 1,415. The boiler pressure was maintained at 210 psi. For the run down the bank into Peterborough steam was shut off and the mid-gear position was selected, no. 2001 reached a top speed of 60 m.p.h. and maintained this easily for most of the way.

Starting from Peterborough station the drawbar pull was 16.3 tons. The cut-off was soon reduced to 12%, but it was raised to 15% near Yaxley and then to 20% at Abbotts Ripton where the line rose at 1 in 200. The speed of the engine at the top of the bank was 45½ m.p.h. From Huntingdon to Biggleswade, despite there being a number of undulations in the gradient profile, the cut-off was 12% for approximately 20 miles before being increased to 18% just prior to reaching Biggleswade. As *Cock o' the North* passed Holme the speed was 62½ m.p.h. and the engine was developing 1,667 drawbar horsepower. Travelling past Arlesey at 67 m.p.h., no. 2001 headed to the 1 in 200 which brought the line to Hitchin. On this ascent 25% cut-off was used and the speed reduced to 61½ m.p.h. with a 4.9-ton pull on the drawbar and 1,800 drawbar horsepower being produced. Between Huntingdon and Hitchin 2 minutes were gained on the scheduled 27 minutes and the total time from Peterborough to this point was 47 minutes 7 seconds. After Hitchin the gradient was again 1 in 200 for approximately 3 miles to Stevenage. The cut-off selected was 30% and the speed was 60 m.p.h. on the bank with a drawbar pull of 5.9 tons

Cock o' the North *has been adorned with an indicating shelter for testing purposes.*

equating to 2,100 drawbar horsepower. Then, *Cock o' the North* was required to slow down to 34 m.p.h. with steam shut off for a permanent way check at Knebworth. For short distances the cut-offs were 40%, then 30% before being shortened to 12% and by Welwyn Garden City the speed had reached 68 m.p.h. At Hatfield cut-off was lengthened to 20% for the 1 in 200 climb out of the town with speed at 60 m.p.h. At the top of the ascent 1,690 drawbar horsepower was recorded in the dynamometer car as the engine was travelling at 63 m.p.h. The 5 miles from Hatfield to Potters Bar were completed in 4 minutes 38 seconds. At New Barnet *Cock o' the North* had reached 76 m.p.h. and steam was now shut off, however, at Wood Green, 4¼ miles away from New Barnet, the speed had only reduced to 73 m.p.h. No. 2001 reached King's Cross 114 minutes 15 seconds after passing Grantham (105½ miles) and 81 minutes 25 seconds from re-starting at Peterborough (76.4 miles).

Driver Peachey also remarks in *A Gresley Anthology* that on the return leg of the test run the coal was in short supply as no thought had been made to replenish the tender when the locomotive stopped at Barkston and the tender was nearly empty at King's Cross. Peachey adds that on the return run the fireman was on the tender getting coal while an inspector was firing the engine, but 'he did not do well and we were losing steam.' 'So, although it meant breaking rule 14, I let the inspector sit on the driving seat to handle the engine while I got the coal down and my fireman looked after the fire. Then I said to the inspector "I'll take charge, the way you're paying the engine we'll

have to stop at Hatfield for coal." Despite these minor problems *Cock o' the North* had produced an impressive performance on both legs of the trial and demonstrated an outstanding ability to climb adverse gradients with heavy loads with an ease that was unprecedented at the time.

Driver Peachey gave up the engine after the dynamometer car test and on the 20th June Driver W.A. Sparshatt took over *Cock o' the North* for the day, working the King's Cross to Peterborough stopping train. The engine returned with the 2.48 p.m. express to King's Cross and managed to gain 5 minutes on the schedule. On the following day no. 2001 was on the 1.30 p.m. to Doncaster and this was notable for a number of instances when the locomotive was pushed to over 80 m.p.h. *Cock o' the North* was nominally allocated to Doncaster shed from new, but as has been detailed, the engine had been working from King's Cross since returning from its public duties in Scotland. It would appear that after the 21st June the locomotive did begin to work from Doncaster shed as on the 27th June no. 2001 underwent its second dynamometer car trial, setting off from Doncaster station to Grantham. The total weight of the train was 637 tons and at Barkston, where the gradients are 1 in 300 and 1 in 200, the highest drawbar horsepower of the trial was produced - 2,070. The dynamometer car was again behind the engine on 2nd, 3rd and 5th July and at the same time the indicating shelter was attached to the front of *Cock o' the North* so that the performance of the cylinders could be examined. No. 2001 worked the 11.04 a.m. to King's Cross from Doncaster and returned on the 4.00 p.m. train on these days. On the 2nd a high of 87½

No. 2001 is seen at King's Cross before a test.

m.p.h. was achieved in the 'up' direction between Little Bytham and Essendine and on the 3rd the top speed was 72 m.p.h. just north of Sandy on the way to King's Cross from Doncaster. On the 5th only one run was possible as no. 2001 had to be taken off the 4.00 p.m. train because of mechanical issues.

On the 10th and 11th July, two more sets of dynamometer car tests were conducted after a change in the blastpipe arrangement had occurred. The diameter of the orifice was enlarged to 5½ in. on the 4th July and on the 11th the taper blocks were changed to no. 1 size. Both combinations had a detrimental effect on *Cock o' the North*'s steaming. From Friday 13th to Thursday 19th July, the tests were focussed on getting data for a variety of the possible blastpipe and chimney settings. These ranged from changing the orifice diameter, altering the taper block sizes, removing the Kylchap cowls and combinations of these. The setting which was deemed to be the best for *Cock o' the North* in service was 5¾ in. diameter blastpipe orifice, no. 3 taper blocks and the complete chimney arrangement.

Before no. 2001 returned to Scotland the engine underwent a number of minor adjustments. *Cock o' the North* was equipped with a Teloc speed recorder and this was driven from the rear coupled wheels. A new valve setting was also applied as the exhaust cam for the left-hand cylinder was altered to close at 50% of the piston's stroke. In addition, the mid-gear cam was changed to allow a small amount of steam to be admitted to the cylinders, while the exhaust valves remained open. A major change was the replacement of the 'sine wave' superheater elements with a 43 element Robinson superheater after problems were experienced with the durability of the former type. The new elements were 1 ft 6 in. longer than the 'sine

wave' elements, 1½ in. diameter and provided a much larger heating surface at 776.5 sq. ft, with the total heating surface increasing to 3,490.5 sq. ft. The increased length also brought the elements to within 9 in. of the firebox tubeplate to increase heat transfer. *Cock o' the North* was in Scotland by the end of July, officially allocated to Edinburgh Haymarket shed on 31st, and was tested by the Scottish area on the 1st and 2nd August. The train weighed 586 tons and was taken to Edinburgh and Aberdeen and back. No. 2001 was then used primarily between Edinburgh and Dundee, leaving the capital in the early afternoon and returning in the early evening. The load limits for the new engine were set at 530 tons northbound and the same southbound, but 550 tons was permissible from Dundee to Edinburgh. These restrictions were put in place because of the limitations of the stations on the route rather than the inability of the engine to haul heavier trains.

Cock o' the North was in Doncaster Works between 24th and 27th August for the centre and right-hand cylinders to receive the altered exhaust cam setting as it had been proved acceptable in service. However, upon inspection of the cams it was found that they were badly worn. This had occurred because the heavy loads experienced by the valves when working against the pressure of the steam was transmitted from the valve spindle to the cams which were in contact through a point. The decision was then taken to replace the scroll cams with the stepped cams and the necessary order for these was placed. While in works no. 2001's tender was fitted with a tablet exchanger on the left-hand side to allow easier working between Montrose and Usan. *Cock o' the North* returned to Scotland and performed three weeks work before the engine was forced to return to Doncaster for a new piston for the left-hand

Cock o' the North's tender is replenished at an unidentified location.

cylinder between 10th and 17th September. After this no. 2001 was back in Scotland until returning south again on 10th October.

During *Cock o' the North*'s first summer in Scotland a journey behind the locomotive was recorded and the details published in C.J. Allen's 'British Locomotive Practice and Performance' feature in *The Railway Magazine* of February 1935. No. 2001 was working from Aberdeen to Dundee in this instance and was coupled to a train weighing 511 tons tare, 550 tons gross; in addition to the passenger carriages there were a number of fish vans, which were attached to the rear of the train. *Cock o' the North* made its way up the acclivity taking the line out of Aberdeen with ease and went past Cove Bay in 9 minutes 23 seconds at a speed of 42 m.p.h. and, subsequently, Stonehaven was passed in 21 minutes 36 seconds. From Stonehaven to Drumlithie summit, 7 miles of very difficult terrain, the time taken was 8 minutes 24 seconds and no. 2001 breasted the top at 44 m.p.h. Laurencekirk, 7.3 miles from Drumlithie, was passed in 6 minutes 54 seconds after some fast running down Marykirk bank, where a top speed of 76 m.p.h. had been reached. The Kinnaber Junction restriction was observed 4 minutes and 10 seconds after Laurencekirk, before the engine reached Montrose, 2.6 miles from Kinnaber, in 5 minutes 51 seconds. The total time for the 40.6 miles between Aberdeen and Montrose was 50 minutes and 1 second - a gain of 8 minutes on the schedule. On the ascent out of Montrose and down to Lunan Bay, *Cock o' the North* took 10 minutes 40 seconds for the 5 miles and reached a speed of 35 m.p.h. The 8.8 miles from here to Arbroath took 9 minutes 43 seconds,

while the time taken from Aberdeen to this point was 70 minutes 24 seconds. On to Dundee, no. 2001 completed the 17-mile section in 22 minutes 7 seconds. In total the 71.3 miles took 92 minutes 31 seconds to travel.

The article also features another record of *Cock o' the North* in service, but the locomotive was lightly loaded with slightly under 400 tons gross, which was also further reduced to under 350 tons at Thornton, and was worked more or less to the timetable. Additionally, Allen mentions that a Pacific locomotive was noted as leaving Edinburgh Waverley station with a train in excess of the prescribed restriction and he elaborates that this was to offer a comparison for coal consumption tests being carried out with *Cock o' the North*. These took place on 1st and 5th October 1934 as no. 2001 had been deemed to be a heavy consumer by the Scottish Area authorities. Doncaster draughtsman E. Windle was sent north to conduct the trials, which took place between Edinburgh Waverley and Dundee. On the 1st, *Cock o' the North* had an empty coal space at Dalmeny, 10 miles from Edinburgh, and the tender had to be replenished by the supply for the station's signal box, leading to a 45 minute delay of the train. In the engine's defence, the coal space had not been adequately filled and the driver not particularly economical in his driving method. A blastpipe alteration occurred for the test on the 5th as the diameter was increased to 6 in. and the smallest taper blocks were fitted. The total amount of coal used by *Cock o' the North* was 5 tons 9 cwt (between Dundee and Edinburgh no. 2001 had a load of 543 tons, reduced by 80 tons at Thornton) or nearly 104 lb per train mile and just over 90 lb per engine mile.

On 6th October 1934, P2/2 Class locomotive no. 2002 *Earl Marischal* entered traffic from Doncaster Works. The engine was similar to *Cock o' the North* in its general design, but the A.C.F.I. feedwater heater and Lentz poppet valves were replaced by a Davies & Metcalfe exhaust steam injector and piston valves. These latter were 9 in. diameter, operated by Walschaerts valve gear with the Gresley conjugated motion for operating the valves for the centre cylinder. The maximum valve travel was 5⅝ in., steam lap (inside cylinder) 1⅝ in. and (outside cylinder) 1⁹⁄₁₀ in. The valves and cylinders were again part of a monobloc casting. Due to the employment of piston valves, the cylinder clearance volumes were significantly reduced, the figures being; 7.74% and 7.19% for the outside cylinders, front and rear respectively, and 7.83% for the front and 7.62% for the back of the centre cylinder. The setting of the Kylchap blastpipe was also different from no. 2001 as the diameter was 5⅞ in. with no. 3 vee bars. However this was subsequently altered to 5¾ in. with the same taper blocks after experience with the engine in service. The alterations to no. 2002 also made the engine weights slightly different, with the total weight in working order being 109 tons 8 cwt, adhesive weight was 80 tons 10 cwt and the maximum axle load was increased to 20 tons 14 cwt. *Earl Marischal* was also equipped with a new type non-corridor tender, which was fabricated using the normal riveting technique, and a reversion was made to cast-disc wheel centres. The tender had the same capacity as no. 2001's, but the weight when it was full increased to 57 tons 18 cwt. Externally the two locomotives had a similar appearance, though, because of the different valve gears used no. 2002's reversing rod was placed underneath the running plate; this latter feature was also different because it curved and straightened at the front of the engine near the smokebox, whereas *Cock o' the North*'s was straight. *Earl Marischal* worked for an extended period from Doncaster shed and during this time it was found that because of the use of piston valves the exhaust was not as vigorous as no. 2001's. As a result the smoke deflectors were not as effective and between 15th March and 17th April 1935 an extra pair of smoke deflectors, similar in shape to the integral type already fitted, were attached to the engine.

During *Earl Marischal*'s time working from Doncaster a journey behind the locomotive was recorded by C.J. Allen and it was featured in *The Railway Magazine* of January 1935. No. 2002 headed the 4.00 p.m. express to Doncaster from King's Cross on a dank night in December 1934 and the train consisted of 17 carriages weighing a total of 580 tons. *Earl Marischal* passed Finsbury Park in 5 minutes 51 seconds, but the locomotive was then restrained to avoid being stopped by signals further along the line. The 1 in 200 rise in the line from Wood Green to Potter's Bar took 11 minutes 36 seconds with the speed falling from 52 m.p.h. to 33 m.p.h. Hatfield was passed at 71½ m.p.h. and *Earl Marischal* had reached this point in 26 minutes 33 seconds after leaving King's Cross, this being only slightly inside the prescribed time. The next 14 miles to Hitchin were scheduled to take 14 minutes and 38 seconds were gained on this, with the speed past here being 70½ m.p.h. High-speed running was maintained through to Huntingdon (a top speed of just under 80 m.p.h. was recorded at Arlesey) and the 26-mile section was completed 3 minutes 12 seconds ahead of time. The total time to Huntingdon was 62 minutes 48 seconds for the 59 miles. The speed was still high on the approach to Peterborough and on arrival at the station 1 minute 52 seconds had been added to the savings on the scheduled time. The number of carriages was reduced before the locomotive continued and the load behind the tender became 400 tons gross.

On the run up to Stoke summit the speed was just under 70 m.p.h. at Tallington, 8 miles from Peterborough, and this point was passed in 10 minutes 47 seconds. At Little Bytham the speed had fallen to just over 60 m.p.h. and a further 6 minutes 46 seconds had elapsed. Passing Corby the speed was 54 m.p.h. and by Stoke summit its was 50 m.p.h. From Tallington to Stoke, the section had been completed in 15 minutes 21 seconds for the 15½ miles. Going down the opposite side of the rise, no. 2002 achieved over 70 m.p.h. at Great Ponton, which was 2 miles from the summit. The total time to Grantham from Peterborough was 31 minutes 35 seconds or 4 minutes 15 seconds early. More high speeds between 70 and 80

P2/2 no. 2002 Earl Marischal *as built in October 1934.*

No. 2002 in mid-1935 after an additional pair of smoke deflectors had been fitted.

m.p.h. were recorded on the falling gradients between Grantham and Newark, but after a level section the line began to rise at 1 in 300 and 1 in 200 for approximately 2½ miles between Crow Park and Askham tunnel. *Earl Marischal* was travelling at almost 70 m.p.h. at the bottom of the slope and nearly 60 m.p.h. at the top. The time from Grantham to Retford was 36 minutes 11 seconds and then on to Doncaster was 19 minutes 47 seconds. From King's Cross the total running time was 167 minutes 39 seconds and the time saved amounted to 12 minutes 21 seconds.

Cock o' the North left Doncaster Works on the 23rd October and further trials were conducted in relation to the optimum setting of the blastpipe. Then, from 1st November to the 24th, no. 2001 was prepared for its planned visit to the locomotive testing station of the French Railways at Vitry-sur-Seine, near Paris. While in works *Cock o' the North* was fitted with stepped cams that provided six cut-off positions; 12, 18, 25, 35, 45 and 75%. In reverse two positions were given; 35% and 75%. After being released from works *Cock o' the North* was lightly worked before it was dispatched to France.

A broadside view of Earl Marischal *with the extra smoke deflectors. The integral pair fitted to the engine originally were reduced in length by 1 ft and the discrepancy between them and the new plates is noticeable here.*

The test plant at Vitry-sur-Seine, Paris, France.

Tests at Vitry

While testing locomotives in service with the dynamometer car could yield useful results, a number of factors, such as the gradient of the land and weather conditions, could distort any data obtained and render it to be of little use to the locomotive engineer. Therefore, it was considered desirable to place a locomotive on rollers in a testing station to conduct tests in a controlled manner and obtain data that was free from unwanted influences. A number of such facilities were in use around the world, but in Britain a modern testing facility did not exist. Locomotive engineers in America were the first to boast the possession of a locomotive testing station, with this being built at Purdue University in the early 1890s, and it was then followed by two more by the turn of the century. In 1905 a testing station was opened by the Pennsylvania Railroad Company at their Altoona Locomotive Works and, as stated formerly, this would later play a pivotal role in the development of their locomotives. In Britain, the G.W.R. opened an experimental station at Swindon Works in the same year as Altoona, but the intention was for it to be used for both testing and the running in of engines that had been repaired in the works. At the time of *Cock o' the North*'s visit to Vitry, the Swindon testing plant was not suited to testing powerful locomotives because the braking system had a limit of only 500 horsepower, although in 1936 the station would be modernised. In Europe, Grunewald was brought into operation in June 1931 for use by the German State Railway and in France, Vitry-sur-Seine testing station was opened in 1933.

During the late 1920s and early 1930s, the person in Britain who was the most vocal about the provision of such a resource was Gresley. He appealed to his peers a number of times through presidential addresses to the Institutions of both Mechanical and Locomotive Engineers, in addition to these being part of technical papers produced by him for the institutions, one being specifically dedicated to the subject of locomotive testing stations. In his presidential address to the Institution of Locomotive Engineers on 28 September 1927, Gresley examined a number of factors that were influencing the construction of locomotives at the time. One point made was that the industry had to keep in touch with the locomotive industries operating overseas or British companies would lose valuable business and one area in which Britain was behind the rest of the world was

the presence of a locomotive experimental station in the country. Gresley comments: 'British locomotive designers and builders work under disadvantageous conditions compared with the engineers of competitive nations. They have not a their disposal any facilities for carrying out experimental scientific research, nor can they obtain the necessary financial assistance to do so.' He goes on to add: 'British railway engineers produce improvements by a sort of slow evolution. New features are tried, and if successful are embodied in new designs. Progress is sure, but it is very slow, and the methods adopted in many cases are empirical rather than scientific.' Further: 'What I suggest is required and is essential if improvements are to be made in steam locomotives, is the provision by the Government of a National Locomotive Testing Plant under the Department of Scientific and Industrial Research and controlled by the Engineering Department of the National Physical Laboratory. Use would be made of such a plant by British Railway companies, consulting engineers of colonial and foreign railways and locomotive builders.' Gresley

A view of the interior of the Vitry testing plant with an Est Railway 4-8-2 locomotive on the rollers.

A close-up of the Est engine's wheels on the rollers.

basis the most promising of the numerous devices which are put forward from time to time for improving the performance of locomotives.'

Subsequently the government formed a committee to look at this matter and Gresley was involved with Sir Henry Fowler of the L.M.S and other leading figures in the railway industry. However, this took place in the early 1930s when the economy was in decline and the idea was dropped by the government, despite being given preliminary approval by the committee. During the time of his involvement with this matter, Gresley delivered a paper on 'Locomotive Experimental Stations' to the Institution of Mechanical Engineers in mid-1931 and in this he explores the features of his 'ideal' testing station. In September 1934, before *Cock o' the North* was sent to Vitry, Gresley delivered a presidential address to the Institution of Locomotive Engineers and again he highlighted an 'urgent need for the provision of facilities for the scientific study of the locomotive in operation.' Gresley went on to praise the French railways and government for working together to produce such a facility at Vitry and bemoaned the lack of action in this country and the ages of the dynamometer cars in use.

The Vitry testing station was in development for several years before it was opened at government expense. Pierre Place designed the facility for the Office Central d'Etudes de Material de Chemins de Fer (O.E.C.M.) and it was put into use in mid-1933 after being erected on land owned by the Chemins de fer de Paris à Orléans; Gresley was in attendance on the opening day. Place was asked by Gresley, after no. 2001's time at Vitry, to produce a paper on the plant and 'Locomotive Test Plants (with Special Reference to the Testing Plant at Vitry)' was delivered to the Institution of Locomotive Engineers in April 1935. Place gives all the mechanical details of Vitry and some of the more pertinent ones are as follows. To prevent vibrations, which could effect the technical readings, the locomotive

points out that a number of similar testing plants have been provided at government expense for the shipbuilding and aerospace industries. His closing remarks were: 'I venture to think that the provision of such a locomotive testing plant would be of national advantage in that it would result in effecting economies in fuel consumption and the operation of railways. It would tend to unification of design with a consequent reduction in the number of types; it would be a means of testing on a strictly impartial and comparative

The rollers at Vitry were controlled by brakes made by the British company Heenan & Froude Ltd.

Train Ferry No. 2.

Cock o' the North *on top of the third row.*

table was fixed to a heavy girder frame and the assembly was mounted to a concrete bed weighing 1,000 tonnes. The locomotive table had 8 rollers and 6 brakes and the former could accommodate up to 12 driving wheels with a maximum axle load of 29.53 tons. However, the table was not limited, and could have been extended by 8 metres if necessary. Self-regulating brakes were used and these were made by Heenan & Froude Ltd. The brakes at Vitry could absorb up to 1,900 horsepower and were able to maintain the locomotive's speed to within 1% of being constant; in comparison with Altoona this was an improvement of the 17% offered by the brakes used there. An Amsler hydraulic dynamometer was employed and was operated through the tractive effort of the engine on test forcing a piston to compress oil in a cylinder. The movement of the oil was transferred to a smaller measuring cylinder that contained a calibrated spring and the deflection of the spring was recorded. Only seven men needed to be employed to reset the locomotive table for a new engine and the process took half a day; if the same locomotive was coupled or uncoupled then the time taken to do this was less than an hour. The top speed an engine on test could reach was 99.4 m.p.h. Up to the time the paper was published 16 engines had been tested at Vitry and had travelled 19,000 miles on the rollers. The total cost for the testing station had been £120,000 and a new dynamometer car had also been provided.

A crowd gathers at Calais to view the new arrival.

The Amsler dynamometer is ready to record the performance of Cock o' the North.

The journey to France began on the night of Tuesday 4th December 1934 as *Cock o' the North* made its way from Doncaster to Parkeston Quay shed and took with it; three forty-ton wagons filled with 'best Yorkshire coal' from Yorkshire Main Colliery, Edlington, near Doncaster, a van containing spare parts and a brake van at the rear of the train. The engine spent the night at Parkeston Quay shed and on the 5th was loaded on to Train Ferry No. 2 for the night crossing to Calais. The ferry was owned by the L.N.E.R., which had been operating train ferries to Holland and the continent through a subsidiary company since 1924 in partnership with the Belgian State Railways, and operated three such vessels in total. The *Harwich and Manningtree Standard* reported a few items of interest: upon the engine's arrival at Harwich and Parkeston large crowds gathered to admire the locomotive; *Cock o' the North* was too long for the turntable so it had to travel to the continent tender first and no. 2001 was also the biggest item to be transported by the ferry. Train Ferry No. 2 docked on the morning of the 6th and after the engine was unloaded and had the fire started, it was moved to Calais shed for the night. On the morning of Friday 7th December *Cock o' the North* began its journey to Vitry with Bulleid, who was with the locomotive throughout the tests, driver G. Trower

and fireman W. Gant on the footplate and arrival at Vitry occurred in the early evening. No. 2001 was run briefly on the locomotive table on the 8th and then preparations were made for the first test on the 13th. The time taken to complete a full set of tests was usually a month and during this time *Cock o' the North* would have the optimum setting of the blastpipe determined before undergoing coal and water consumption tests.

No. 2001 was run up to 76 m.p.h. on its first test at Vitry and at this time the engine was filmed for a British Pathé newsreel, which also documented the locomotive's crossing to France. Unfortunately, no. 2001's driving axlebox on the right-hand side ran hot and the tests had to be suspended for the driving wheels to be dropped and the axleboxes examined. This occurred on the 14th at the nearby Paris-Orléans Railway electric locomotive works and it was found that both boxes had to be replaced. *Cock o' the North* was ready to return to the test bed on the evening of the 16th to be run in before the trials were to begin again on Monday 17th December. The day's testing was conducted at 60 km/h (37.2 m.p.h.) using 12, 18, 25 and 35% cut-off positions, but at the short cut-offs a large amount of smoke was emitted from the chimney. On the 18th, the same speed was to be used and longer cut-offs

The view from the floor with no. 2001 on the rollers. Note the Teloc speed recorder has been fitted.

tested. However, no. 2001 was only briefly on the rollers before the crank pin on the left-hand side overheated and the rest of the day was lost to replacing the bush. After a period of light activity on the 19th, the tests commenced again, but only a short time later the coupling rod bush was hot. The Paris-Orléans Works again had to be used to remetal the coupling rod as the spare bush had been used the previous day. Problems were then encountered with the quality of the remetalling, as this was deemed to not be of the required standard, and the work on the coupling rod was not accepted until the 20th. *Cock o' the North* produced a full day's test on the 21st and after being bedded in, ran continuously at 80 km/h (49.7 m.p.h.), 100 km/h (62.1 m.p.h.) and 120 km/h (74.5 m.p.h.) at 12% cut-off. After the completion of this series the tests were concluded until after the Christmas period.

A valve examination took place on New Year's Day and then on the 2nd January 1935 the tests were resumed. The locomotive was run at 100 km/h and 18% cut-off for less than half an hour before the test was stopped due to a mechanical breakage on the testing table. Prior to the trial commencing the ashpan had been modified in an attempt to increase the air flow to the fire and to reduce the smoke being produced at the chimney. However, this modification effected no change and in light of the stoppage the decision was taken to change the blastpipe tops to 5½ in. diameter with no. 3 vee bars. After the trials commenced on the 3rd it was found that the alteration of the blastpipes had been successful in its aim, however, a large amount of coal was now being ejected from the chimney. *Cock o' the North* was being run at 80, 100 and 120 km/h and 12, 18, 25 and 35% cut-off when the axlebox of the first pair of coupled

Cock o' the North *at speed.*

Another view of no. 2001 running on the locomotive table.

wheels was found to be overheating and testing stopped while the necessary repairs were completed. Three days passed before no. 2001 was ready and during this time the cams were examined, repaired because of some wear and reassembled. The blastpipe tops were also changed to 5¾ in. diameter with no. 0 taper blocks. When *Cock o' the* *North* returned to the testing station it was run at 60 km/h with the 12 and 18% cut-off positions before the axleboxes again gave trouble. At this point Gresley instructed that the lubricating oil for the axles should be of the same type used by the French locomotives. On 10th January, no. 2001 was worked on the rollers for several hours at low

No. 2001 was filmed by British Pathé while at Vitry in addition to the photographs documenting the event.

speeds and remained cold in the axleboxes. During the day the locomotive was seen in action by the Committee for Scientific and Industrial Research. Trials were resumed on the 12th January at speeds of 60, 80 and 100 km/h. At the latter speed axlebox heating again occurred and the order was given to remetal the box with a complete white metal surface in accordance with the standard French practice. The 14th was dedicated to running the engine lightly in order to run in the axlebox sufficiently. In the meantime, the blastpipe tops had reverted to 6 in. diameter, but with no. 1 vee bars. On the 15th the tests were conducted at 60 km/h and at the 12, 18 and 25% cut-off positions and then on the 16th the speed was increased to 80 km/h with the same cut-offs. At the speed and 25% cut-off the horsepower of the locomotive was indicated to be 2,278. Subsequently, 100 km/h and 12% cut-off was tested, but during this an axlebox failed again and the decision was made for all the axleboxes to be remetalled in the French custom and nine days were lost to this; the lubricating oil was again under close scrutiny and there was a change of the type used. *Cock o' the North* was run briefly at Vitry on the 25th before travelling on the line between Vitry and Etampes at 25 m.p.h. to test the axleboxes and they were found to be operating satisfactorily.

The tests were then moved to the next stage of recording coal and water consumption, despite the best blastpipe setting not being found, it will be noted. However, only one test at 60 km/h and 12% cut-off was possible before the right-hand driving axlebox was overheating. The offending box was remetalled over the next two days and then *Cock o' the North* was moved to Tours Works for a full inspection, but on arrival Bulleid ordered that no further action would

be taken because the axleboxes were showing no signs of heating. High-speed running tests were then scheduled to take place between Tours and Orléans (72 miles) on 1st February. Two journeys there and back took place. On the first to Orléans the average speed was 52 m.p.h. and to Tours it was 61 m.p.h., while the top speed attained was 78 m.p.h. During the second trip the average speeds were 64 m.p.h. and 63 m.p.h.; the highest speed was 82 m.p.h. The successful running of *Cock o' the North* in these instances led to tests with three counter pressure locomotives being authorised for the 4th, 5th and 6th February with the Paris-Orléans dynamometer car. In this method of testing steam entered the blastpipe of the counter pressure engine before being admitted to the cylinders and this offered resistance for the engine under test and kept it running at a constant speed. The trials occurred between Saint-Pierre-des-Corps and Orléans, which was a distance of approximately 65 miles. *Locomotives of the L.N.E.R. Part 6B* (1991) produces a table of the results obtained on these trials. On the 4th, 63.7 miles were travelled with the regulator open and the average speed was 44.7 m.p.h. and the cut-off was 18%. *Cock o' the North* produced an average drawbar horsepower of 1,236 and the firing rate was 56 lb per minute. Travelling in the opposite direction on the 5th, the cut-off was increased to 25% and the average speed rose to just under 56 m.p.h. The average drawbar horsepower was 1,587 and the firing rate 88 lb per minute. No. 2001 travelled for a slightly reduced distance with the regulator open on the 6th as the boiler pressure could not be maintained anywhere near the 200 psi required and was in fact, on average, 169 psi. This was also 30 and 20 psi lower than had been maintained, on average, during the previous two days of testing. The cut-off on the 6th was 35% and the average speed was 68 m.p.h. The average drawbar horsepower was 1,883 and firing rate 107 lb per minute. For 35 minutes of the test 1,910 drawbar horsepower was recorded continually. A problem encountered was that the sharp blast was again causing coal to be ejected from the chimney.

Cock o' the North returned to the test plant on the 8th February and it was possible to run the engine all day without the axleboxes overheating. The test was performed at 90 km/h (55.9 m.p.h.) and 18% cut-off with a partially opened regulator. The boiler pressure stuck close to 220 psi and the steam pressure was 125 psi. When the test was completed the regulator was opened fully for a time, with the test speed and cut-off remaining the same, and 1,600 horsepower was recorded. On the 9th the intention was to return to this horsepower and speed, but lengthening the cut-off to 25%. This was not possible, however, because the necessary boiler pressure could not be maintained and, despite attempts to save the test, it was ultimately abandoned and *Cock o' the North* left the table. On the 10th the boiler was washed out and thoroughly cleaned after a number of the tubes were found to be partially blocked by detritus. The blastpipe orifice diameter was also made 6 in. with no. 3 vee bars. On the 11th the test conditions from the 9th were resumed and the boiler was worked satisfactorily for three-quarters of the test before the steaming problems resumed and the test was discontinued just before the completion time of

two hours, having worked on average at a good deal less than the required horsepower. During the trial, difficulties had also been experienced with the wheels running on the rollers incorrectly and there being some movement sideways, this resulting in a warm axlebox which cooled after a break was taken. The test was repeated after the pause in activity, however, after a quarter of the test the leading axlebox was hot and the decision was taken to abandon the trials completely.

Arrangements were then made for *Cock o' the North* to be exhibited at the Gare du Nord, Paris, on the 17th February before it returned to England. The engine was prepared for its engagement at the Nord Works at La Chapelle, Paris, on the 15th and 16th. On the 14th, on its way to the works, no. 2001 was relieved of the contents of two of the coal wagons which had not been used and this was given to the P.L.M. for it to be used on tests with one of their engines. No. 2001 was on display to the public at the Gare du Nord's platform no. 1 from 11.30 a.m. to 7.00 p.m. and a contemporary report notes that over 5,000 people viewed the engine and a Chemins de fer du Nord saloon carriage and one of the company's 'Super' Pacifics which had joined *Cock o' the North* for this brief event. Later in the day the exhibits were joined by the new Nord diesel-electric streamlined train, which had a top speed of 80 m.p.h. and could accommodate 144 passengers. *Cock o' the North* left for Calais with its train on the 20th and returned to England on the night ferry of the 21st February.

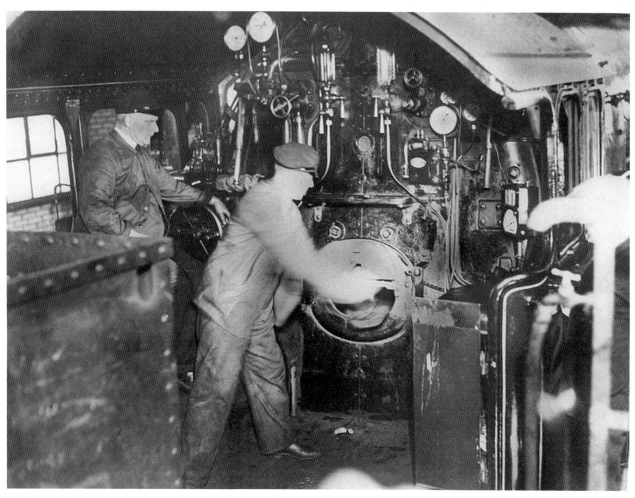

Fireman Gant hard at work while driver Trower looks on. Both men were from Doncaster shed.

Cock o' the North *on exhibition at the Gare du Nord, Paris. Approximately 5,000 inspected the engine.*

Out of nine possible days testing before the Christmas break *Cock o' the North* had lost seven to mechanical problems and consequent running in of the parts before testing could be performed. Of the 36 available in the new year approximately 23½ days were lost for similar reasons, while the tests on the track took up a further seven in total. *Express Steam Locomotive Development in Great Britain and France* by Colonel H.C.B. Rogers (1990) explains that the French engineers informed Bulleid that the problems experienced by *Cock o' the North*, with regards the heating of axleboxes, which up to the time had not occurred on the engine when in service in Britain, had also afflicted other engines and this was due to the rigidity of the test bed. Further, when Gresley went to visit Chapelon for his advice on no. 2001's problems, the response was that trials with the counter pressure locomotives could be more fruitful.

In the discussion of Spencer's 'The Development of L.N.E.R. Locomotive Design 1923-1941' Bulleid commented: 'when tested on the open road between Tours and Orléans she developed a very high horsepower, of the order of 2,800, and again showed herself to be an efficient engine from the point of view of coal consumed per d.b.h.p.' O.S. Nock in 'British Locomotive Practice and Performance' in *The Railway Magazine* of May 1960 records these figures as 2.61 lb at 44 m.p.h., 3.1 lb at 55 m.p.h. and 3.25 lb at 68.5 m.p.h. Rogers (1990) also reproduces the relevant figures for a Paris-Orléans Chapelon

Pacific at 68 m.p.h. When the drawbar horsepower was 1,900 the coal consumption per drawbar horsepower hour was 2.31 lb. At 56 m.p.h. 2,700 drawbar horsepower was developed and 2.68 lb of coal was used per drawbar horsepower hour. Whereas the 1,910 drawbar horsepower produced by *Cock o' the North* lasted for 35 minutes, the Paris-Orléans Pacific maintained this figure for the duration of the test. It will also be noted that Bulleid's assertion that no. 2001 produced 2,800 drawbar horsepower has not been corroborated and has been challenged by a number of people, even though he was adamant that the figure was in fact obtained. In relation to *Cock o' the North*'s performance on the testing bench Bulleid commented in Spencer (1947): 'She was not an extravagant engine at all, but was in fact extremely efficient on the testing plant, and compared favourably with the French engines in her coal consumption per rail-h.p.' One area that no. 2001 was different to the French engines was the production of smoke from the chimney. Bulleid thought that this was because of an inadequate flow of air to the grate and resulted in there being incomplete combustion in the firebox. Also, on the road tests the French footplatemen accompanying the English crew suggested an enlargement of the firebox door could have been beneficial, but this was not carried out subsequently. Gresley commented on the tests to the Institution of Mechanical Engineers when he made his presidential address, nearly two years after their completion, in 1936: '...I sent one of our new engines to the experimental station of the French railways at Vitry, near Paris, and obtained most valuable information from the research then carried out, which enabled me to improve the design and efficiency of engines built to that and other types.'

Over three months elapsed before *Cock o' the North* was ready to enter traffic in Scotland again and during this time a few modifications were made to the engine as a result of the trials in France. The grate was altered to allow more air to flow between the fire bars and the the original 30% was increased by 26%; W1 no. 10000 was also modified in this manner in May 1935. While on test in the mid-December 1934, it was noticed that *Cock o' the North's* exhaust beats were uneven. The Dabeg Company, holders of the Lentz patent in France, inspected the valve chests on 1 January 1935 and suggested that the valve setting should be made to crank angles rather than valve openings, but this does not appear to have been implemented. Dabeg also recommended the use of bronze pins in the cam boxes, as was the practice on the continent, when they were checked on 5th January and even presented the team with a set for use, but it is again unclear if they were fitted. The uneven beat was eventually attributed by the L.N.E.R. to the difference in the clearance volumes between the cylinders and after no. 2001's return an attempt was made to reduce these. They became; 10.88% and 10.21% for the outside cylinders, front and back respectively and the inside; 14.35% for the front of the cylinder and 14.05% at the rear. The heat generated in the cam boxes was also

seen to be detrimental to the operation of the cams and this time out of traffic was taken to install an arrangement to cool the oil in the cam box. A pump for each cam box took the oil to copper pipes fixed to a frame mounted in front of the cylinder casting.

After these modifications were carried out between 23 February and 30 March *Cock o' the North* again worked for a period between Doncaster and King's Cross and was also exhibited at Stratford Works in early May. Later in the month no. 2001 was again in Doncaster Works to be prepared to return to Scotland and was subsequently dispatched in early June. *Earl Marischal* was also moved to Haymarket shed in early June after it was given attention at Doncaster Works, but it was transferred to Dundee shed on the 22 June and began operating between Dundee and Aberdeen. *Cock o' the North* was generally limited to working from Edinburgh to Dundee and was first noted in Tay Bridge station by the *Dundee Courier* on Wednesday, 12th June.

A4 Class and V2 Class

From *Cock o' the North's* completion up to the two P2 Class locomotives' introduction into regular service in Scotland, Gresley was involved with another important project, namely the introduction of a high-speed service from King's Cross to Newcastle. He had been inspired to do this from events on the continent and America where a number of similar services were running between important population centres. In particular, the Deutsche

No. 2001 pictured at Edinburgh Waverley station.

Reichsbahn-Gesellschaft diesel electric Class SVT877 locomotive running between Berlin and Hamburg, known as the Fliegender Hamburger or 'Flying Hamburger', and the Bugatti railcars working in France, both being introduced in 1933. After deciding that steam power could produce a better performance with a longer train than formed these services, Gresley had A1 no. 4472 *Flying Scotsman* trialed between King's Cross and Leeds to see if it was possible in reality. On 30th November 1934 no. 4472 was coupled to a train of four carriages at King's Cross and this was increased to six at Leeds. *Flying Scotsman* proceeded to produce superlative performances on both journeys and the plans were advanced with. In late April 1934 the first outline drawing for the proposed new locomotive for the service had appeared and it resembled no. 2001 very closely, with rotary cam poppet valves, diagram 94A boiler working at 220 psi and similarly streamlined front end, but with a Pacific wheel arrangement (6 ft 8 in. diameter coupled wheels) and a normal chimney. On 5th March 1935, A3 Class Pacific no. 2750 *Papyrus* was worked with a six carriage train, including the dynamometer car, from King's Cross to Newcastle and back, and reached 108 m.p.h. on the return leg, averaging close to 70 m.p.h. over the complete journey. Presumably, as this coincided with *Cock o' the North*'s return from Vitry, a number of alterations were made to the design of the proposed engine before the end of March. The poppet valves were abandoned in favour of piston valves with Walschaerts and the Gresley conjugated motion, the length between the boiler tubeplates was shortened by 1 ft to 17 ft 11¾ in. and this led to an increase in the length of the firebox combustion chamber to 4 ft; when built the boiler was classified diagram 107. The smoke deflectors were still similar to the two P2s, but the boiler cladding was brought down over the coupled wheels. The next drawing appeared

A4 Pacific no. 2509 Silver Link *in Doncaster Works' Weigh House.*

in the following month. In the meantime wind tunnel experiments had been conducted to determine the best form of the streamline casing and while the drawing featured a curved running plate the front end had not been altered. This was added subsequently and was inspired by the Bugatti railcars as it was found that the shape provided the best method of dispersing the exhaust emitted from the chimney. An important step forward was the raising of the working pressure of the boiler to 250 psi to allow a greater reserve of steam for fast uphill running. Construction of

Gresley's experience with the P2s led to the abandonment of poppet valves and the A.C.F.I. feedwater heater and the implementation of an improved method of smoke deflection when the design for the A4 Class was produced.

2-6-2 no. 4771 Green Arrow *was the first of the V2 Class, which eventually numbered 184 engines.*

the first four engines of the new A4 Class did not start until the end of June 1935. The locomotives were to also feature carefully designed steam passages with large valves, following on from the practice employed on the P2s. The piston valves were 9 in. diameter and particular care was taken in designing the port openings, making the steam passages to the cylinders straight and making the finish of the steam passages as smooth as possible to encourage free steam flow. Other features in common with the P2s were the v-shaped cab front and the Crosby tri-note whistle, but the notable difference was the absence of the Kylchap double blastpipe and chimney. The single chimney was of the 'jumper top' type employed by Churchward on the G.W.R. This consisted of a cap attached to the blastpipe which lifted when necessary to increase the area available for the exhaust steam to escape. The first A4, no. 2509 *Silver Link*, entered traffic on 7th September 1935 and after running in was demonstrated to the press on 27th September between King's Cross and Barkston. The train was of several coaches that had been especially designed and built for the new service, which was to be titled the 'Silver Jubilee', and this weighed 220 tons. From Stevenage to Offord, 25 miles, the speed was over 100 m.p.h. and a top speed of 112 m.p.h. was attained. The three other A4s, nos 2510-2512, had followed *Silver Link* into service by the end of the year.

Another design coming to fruition during this period was for a new high-speed mixed traffic locomotive, later to be classified V2. The first diagram for this class was dated August 1934 and, as with the A4 Class, there was a striking resemblance to no. 2001 with all the main features included. However, by the following August this had evolved to a 2-6-2 version of the A4 Pacifics with the same length firebox combustion chamber. When the first locomotive, no. 4771 *Green Arrow*, appeared in June 1936 the combustion chamber had reverted to 3 ft long, but the boiler size had increased to 17 ft between the tubeplates.

Some notable features surviving from the original drawing were the v-shaped cab front, boiler pressure of 220 psi, and 6 ft 2 in. diameter coupled wheels. As the class multiplied, engines were allocated to Haymarket, Dundee and Aberdeen to work express passenger and goods services; in the case of the former services the limit to the load was the same as the A3 Class Pacifics.

Working in Scotland and Four New P2s

During the P2s' first summer in service in Scotland C.J. Allen and O.S. Nock were on hand to record a number of performances by the locomotives for *The Railway Magazine* and the particulars of these appeared in 'British Locomotive Practice and Performance' of January 1936. *Cock o' the North* was recorded three times hauling the 'Aberdonian' from Dundee to Edinburgh. This service was heavily loaded on all occasions and the train also included several fish vans coupled to the rear. The total loads for the journeys were 510, 530 and 540 tons gross. The lightest load was handled adequately by the engine and after losing some time at the start of the journey there was no real effort made to drive the engine faster than necessary. On the banks the loss of speed between the bottom and the top was noticeable, with a particular instance being from Ladybank to Lochmuir. The arrival at Edinburgh Waverley was just over the scheduled time. With the heaviest load *Cock o' the North* lost as much as 3 minutes to Leuchars Junction, but this was gradually made up through some quick uphill running and fast speeds downhill. For example, the four miles beginning at Ladybank and ending at Lochmuir took 5½ minutes and down to Burntisland just over 70 m.p.h. was attained. The completion time of this journey was only 22 seconds outside the 85-minute schedule. O.S. Nock recorded the other train and for this he had the privilege of being present on the footplate, affording

Gresley's 'big engines' pose for a publicity photograph at Haymarket Shed. The A3 is no. 2796 Spearmint.

him the opportunity to take details of the working of the engine. The general characteristics of this was that the regulator was not fully open for any part of the journey and the steam was only allowed to pass through either a three-fifths or three-quarters opening, but on a number of occasions, especially when running downhill, steam was shut off. Apart from when starting and increasing speed, the cut-off used was 18%. The pressure drop between the boiler and the steam chest was between 3 and 15 psi, although there were two occasions noted when they were the same. There was a general gain of time throughout the running of the train and arrival occurred 3 seconds under time.

Cock o' the North *at Inverkeithing with an Aberdeen express consisting of nine carriages.*

Earl Marischal's three records were taken between Aberdeen and Dundee and two of the trains were the 'Aberdonian', weighing 510 tons and 530 tons, while the other was the 10.20 a.m. express which weighed 515 tons gross. O.S. Nock was again on the footplate to record the particulars of this latter journey and these contrasted with *Cock o' the North*. Full regulator was used for large sections and was never fully closed, but was reduced to the smallest openings when necessary. The pressure drop between the boiler and the steam chest when the regulator was fully open was nil. A greater range of cut-offs were used and the selection was generally 25% when working at full regulator. With regards the journey, *Earl Marischal* reached Cove Bay 17 seconds slower than *Cock o' the North* had the previous summer and was also moving at a slower speed. The time to Montrose was also 5 minutes slower, but the engine was still ahead of schedule and at the completion of the service the saving was nearly 4 minutes. Similar times were made with the other two trains and the heaviest of these brought a particularly good performance from *Earl Marischal*. After losing time progressing out of Aberdeen, no. 2002 went on to gain 4 minutes and the run out of Montrose was quite noteworthy. On the 1 in 88½ and 1 in 111 gradients the locomotive accelerated from 26 m.p.h. to 35 m.p.h. and the time to the summit was the quickest of the three.

Cock o' the North and *Earl Marischal* continued to work their respective duties throughout the winter of 1935/1936 and up to the spring. At this time they were soon to be joined by four new P2 Class members, which were to have been built with no. 2002, but the L.N.E.R. put them on hold to await the outcome of the first two engines' performance in service. As *Cock o' the North* had not distinguished itself sufficiently from *Earl Marischal*

to warrant the perpetuation of its distinguishing features, the new engines were to be the same mechanically as the latter engine. However, one important difference was the addition of the wedge-shaped front end, which had been applied to the A4 Class, and was necessary because of the difficulties of dispersing the exhaust experienced with the original design as fitted to no. 2002. The first engine to exit Doncaster Works was no. 2003 *Lord President* on 13th June 1936. Despite the addition of the front end, the weight of the engine was reduced to 107 tons 3 cwt and the axle loads were now, front to rear; 10 tons 14 cwt, 19 tons 17 cwt, 20 tons, 19 tons 10 cwt, 19 tons 12 cwt, and 17 tons 10 cwt. The adhesive weight became 78 tons 19 cwt. *Lord President* was given a streamlined non-corridor tender, no. 5576, which had been developed for use on the A3 and A4 Class Pacifics. The side sheets were slightly higher, 7 ft 5½ in., from those used on the new type non-corridor tenders fitted to nos 2001 and 2002. The division plate was also raised to 8 ft 3½ in. and the coal bunker and water filling space were also partially covered to form a streamlined appearance. The water capacity was 5,000 gallons and the coal capacity was 8 tons with the total weight of the tender being 60 tons 7 cwt. The coal capacity was later altered to 9 tons as the covering was removed on all the streamlined non-corridor tenders because, especially with the A4 Class on the non-stop services, problems were encountered with locomotives depleting their coal supply. No. 2003 also featured a change to the cab profile as the cut out was distinctly shorter because of the introduction in June 1935 of rear portions to the driver and fireman's seats. To improve conditions further a rubber sheet was provided between the cab roof and tender front to keep the weather at bay. After being briefly run in, *Lord President*

P2/2 no. 2003 Lord President *was the first of four new engines completed as part of order no. 334 in 1936.*

The frames of no. 2006 Wolf of Badenoch *have been married with the cylinders.*

was allocated to Haymarket shed.

On 11 July 1936 no. 2004 *Mons Meg* entered traffic, but it did not go straight to its intended home at Haymarket, as it encountered a mechanical issue after being sent into traffic when completed and time was lost to repairs. As a result, *Mons Meg* did not arrive in Edinburgh until the end of August. No. 2005 *Thane of Fife* had been sent to

Dundee shed in the meantime, having been completed in mid-August. Nos 2004 and 2005 were equipped with two different chimney arrangements when entering traffic. *Mons Meg* had a 5 in. diameter pipe siphoning exhaust steam from below the blastpipe tops, taking it to an orifice behind the chimney, and this was initially worked at the discretion of the driver. This arrangement was deemed

Work progresses on no. 2004 Mons Meg.

P2/3 no. 2006 entered service in September 1936 carrying diagram 108 boiler no. 8934 which incorporated a longer combustion chamber and reduced distance between the tube-plates. The difference between the two types used by the class was visible on the number of washout points on the side of the firebox above the handrail. The other five engines had five on the left, one hidden by the cab, and four on the right, while Wolf of Badenoch possessed five on each side with one obscured by the cab on the right side. The picture was taken in October 1936.

The diagram 108 boiler in June 1936 before being fitted.

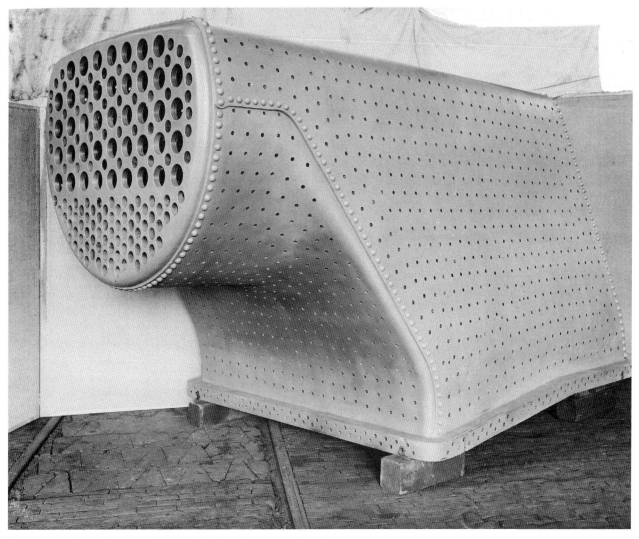

The copper firebox in Doncaster Works' Boiler Shop during April 1936.

No. 2001 Cock o' the North as rebuilt with piston valves, exhaust steam injector and streamlined front end in April 1938. Note also the alteration to the cab side sheets and the hole at the front of the tender for tablet exchange apparatus.

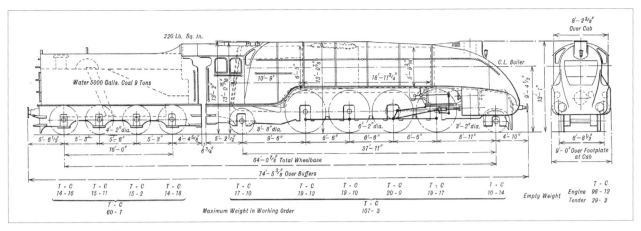

Engine diagram for P2/2 Class nos 2003-2005.

necessary because a disadvantage of the Kylchap system as applied to the class was a tendency for the sharp blast action to pull the fire when the P2s were working near the maximum cut-off position. *Thane of Fife* was equipped with a single blastpipe and chimney and this was 6 in. diameter, but it was later replaced by the 'jumper top' pattern as used by the A4 Pacifics.

The last P2 engine to be erected at Doncaster was no. 2006 *Wolf of Badenoch* and it left the works at the beginning of September. The notable feature of this locomotive was that it was equipped with an altered design of boiler, which was similar to the A4 Class diagram 107 with a longer combustion chamber, although it retained the 50 sq. ft grate. The heating surface of the firebox was increased to 252.5 sq. ft, while in the tubes and flues it was reduced to 1,281.4 sq. ft and 1,063.7 sq. ft respectively. The superheater heating surface was 748.9 sq. ft and the total heating surface

of the boiler was 3,346.5 sq. ft. This boiler was classified diagram 108. The four new P2s again featured names with a connection to Scotland and a competition had been staged for The Boy Scouts to suggest names for the locomotives.

Wolf of Badenoch was allocated to Haymarket when new, but had been moved to Aberdeen by November, where it joined *Earl Marischal*, which had been transferred from Dundee in September. This move was undertaken while no. 2002 was in Doncaster Works undergoing its first general repair, which took place between early August and mid-October 1936. During the course of the repair, the opportunity was taken to rebuild the front end of the locomotive with the front end streamline casing. *Cock o' the North* had been in Doncaster for its first general repair between 27 May and 7 July 1936, although the modification made to *Earl Marischal* was not applied in this instance. The decision regarding a change was not made until the latter

No. 2003 Lord President *stands outside Dundee Tay Bridge shed. The locomotive was allocated to the depot between September 1936 and October 1942.*

No. 2002 Earl Marischal *with streamlined front end, October 1936.*

half of 1937 when no. 2001 arrived at Doncaster Works to receive not only the new front end, but to be fitted with piston valves and an exhaust steam injector. Further, the setting of the Kylchap blastpipe was changed to follow that applied to *Earl Marischal* and the other P2s.

As applied to no. 2001 the poppet valves had been a disappointment because the large clearance volumes necessary resulted in an excessive waste of steam and therefore increased the coal consumption. The replacement of the scroll cams with infinitely variable cut-off to stepped cams with six positions of cut-off had also been a major setback as it precluded working the engine at its optimum levels. Criticisms of the valve gear in this form (much the same as those aimed at the C7 Atlantic stated previously)

were that the selections were not suitable as they either gave too little or too much power than necessary. Bulleid, in 'Locomotives I Have Known' (1945), comments that excessive wear with the stepped cams was also a problem and contributed to the decision to convert. The A.C.F.I. feedwater heater was also noted on several occasions to be functioning incorrectly in service, during some dynamometer car trials and at Vitry. The method of working *Cock o' the North* over the Edinburgh to Aberdeen line meant that the desirability to have the regulator open for the majority of the journey was reduced so exhaust steam was not available for periods to heat the feedwater. *Cock o' the North* returned to Haymarket in its altered state, and reclassified P2/2, during April 1938.

No. 10000 was also fitted with a Kylchap double blastpipe and chimney and developed a similar problem as no. 2002 with regards exhaust drifting leading to this modification to the deflectors.

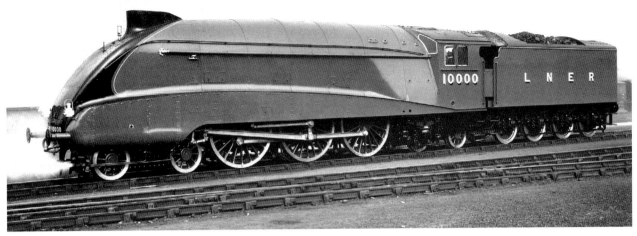

Gresley admitted defeat with the W1 project and no. 10000 was rebuilt with streamlined casing and diagram 111 boiler which had the same dimensions as the diagram 108, but working at 250 psi.

Speaking in the discussion of Spencer (1947) Bulleid suggests that the high coal consumption of *Cock o' the North* was the result of misuse of the engine by the Scottish Area rather than any other major fault with the design. He articulates that instead of working heavy trains for long distances, for example between Edinburgh and Aberdeen, no. 2001 was used to Dundee with a lighter load that was perhaps suitable for and engine of its size. *Cock o' the North* would then be on shed for longer than desirable waiting for the return service with the consequent waste of fuel. The combination of the poppet valves and Kylchap blastpipe and chimney would also appear to have had a detrimental effect on the coal consumption as the rapid opening of the poppet valves and the draught created by the chimney arrangement had a tendency to pull holes in the fire. To reduce the problem a thicker fire would have been necessary and resulted in greater use of fuel. In relation to the P2 Class as a whole, Hughes in *Sir Nigel Gresley: The Engineer and His Family* (2001) records the average figures of the locomotives' coal consumption for the years 1936-1939 as 73 lb per mile. In relation to the A3 Class Pacifics this was 14 lb more, but to be taken into account is the larger grate area of the P2s and their handling of, generally, heavier loads on adverse gradients.

During 1935 W1 no. 10000 had been fitted with a Kylchap double blastpipe and chimney arrangement which was designed specifically to suit the locomotive. The blastpipe tops were smaller than those used with nos 2001 and 2002 at just under 5 in. diameter and no. 10000 was tested for a time with the L.N.E.R. counter pressure locomotive, B13 4-6-0 no. 761, plus the dynamometer car to determine the best combination of vee bars with the blastpipe top. Although the optimum setting was not arrived at, a satisfactory permutation was reached, but this necessitated additions to the smoke deflectors. A3 Class no. 2751 *Humorist* was the next engine to be equipped with the type of chimney in mid-1937 and it

was developed from that fitted to no. 10000 (it may be noted *Humorist* was involved in a number of experiments involving smoke deflectors prior to this). The size of each of the blastpipe tops was increased to 5¼ in. diameter with no. 3 taper blocks and the chimney dimensions were also enlarged to 1 ft 5⅜ in. bottom diameter, 1 ft 3 in. at the choke and 1 ft 4¹³⁄₁₆ in. at the top. This had evidently proved successful in a short time as A4 no. 4468 *Mallard* was the first of the class to obtain the Kylchap arrangement before it entered traffic in March 1938, but the diameter of the blastpipe tops was reduced to 5 in. Four months later no. 4468 set the world speed record for steam locomotive as it achieved close to 126 m.p.h. on the descent from Stoke signal box. The final three engines of the A4 class, nos 4901-4903, also conformed in being fitted with the Kylchap exhaust. Incidentally, around the time of *Humorist*'s fitting of the Kylchap arrangement the L.M.S. also equipped a Stanier 'Jubilee' Class 4-6-0. This was no. 5684 *Jutland* and after trials were conducted, during which tearing of the fire and the ejection of coal from the chimney had been noted, it was removed from the engine.

The P2 Class went about their duties in a quiet fashion into the early 1940s. In October 1942 the allocations were slightly changed, with *Lord President* and *Wolf of Badenoch* being moved to Haymarket shed and leaving *Earl Marischal* and *Thane of Fife* the sole representatives of the class at Aberdeen and Dundee respectively. Only one more footplate run was recorded behind a member of the class, no. 2004 *Mons Meg*, and this was by E.H. Livesay, appearing in T*he Engineer* of October 1939. The train was a light 'Aberdonian', 320 tons gross, and the few points of note were the high downhill speeds and the 'easy' riding of the engine, which has also been noted by several others in reference to the class in general. After arrival at Dundee, *Mons Meg* was on shed for three hours before returning to the capital with the 8.20 a.m. service and this weighed 360 tons. Speeds up to 60 m.p.h. were attained, but the arrival was six minutes late as the engine encountered a number of signal checks, in addition to the stops at stations. The hard work of the fireman was also noted. No. 2004 was then returned to Haymarket where it awaited the next booked duty to Dundee, which was at 2.00 p.m. On this occasion 1 minute was gained on the schedule and the average speed of the run with the 355-ton train had been 40 m.p.h. The locomotive was worked on these occasions with partially opened regulator and longer cut-offs.

Fireman Bray, driver Duddington and inspector Jenkins pose after no. 4468 attained the world speed record for a steam locomotive.

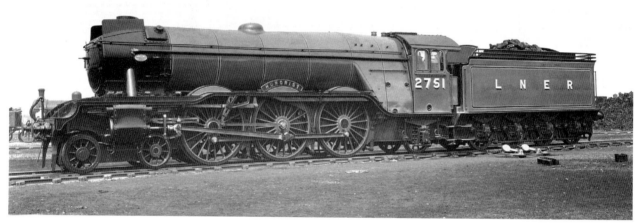

A3 no. 2751 Humorist *with one of the many arrangements attempting to solve the smoke deflection problem; this one dates from April 1933.*

No. 4903 Peregrine *was the last A4 to be fitted with the Kylchap double chimney until a start was made on the remainder of the class in May 1957.*

Chapter Three **Rebuilding**

Thompson Takes Over

Sir Nigel Gresley died in office on 5th April 1941 after a brief illness and this sad event left the L.N.E.R. Board with the hard task of naming a successor to the company's C.M.E. of 18 years. The suggestion had been made that overtures were made to Bulleid, who had left the L.N.E.R. to become C.M.E. of the S.R. in 1937, and R.C. Bond, assistant to L.M.S. C.M.E. William Stanier. However, by the 28 April 1941 Edward Thompson was named the successor to Gresley. He was one of the most senior mechanical engineers with the company, having been with the L.N.E.R. since Grouping and had formerly held positions with the N.E.R. and G.N.R. Interestingly, Thompson had succeeded Gresley once previously in 1912 when he became Carriage and Wagon Superintendent for the G.N.R.

Upon Grouping, which Thompson saw in from the same position on the N.E.R., having returned to the company in 1921 (he was son-in-law to C.M.E. Sir Vincent Raven), Thompson became Carriage and Wagon Engineer for the N.E. Area of the L.N.E.R. While involved with this side of

Gresley on his way to receive a knighthood in 1936.

Edward Thompson (1881-1954) C.M.E. 1941 to 1946.

rolling stock, he has not been noted for any design work in particular, but it has been said that he was quite adept at organisation and management matters. Thompson's next promotion took him to Stratford Works in 1927 as Assistant Mechanical Engineer, then in 1930 he became

Mechanical Engineer, Stratford. While in this position Thompson made an experiment of rebuilding a B12 Class 4-6-0 that was under his authority at the works. As stated before the B12/2s were modified to feature Lentz poppet valves, which were not wholly successful, and at the same time the original B12s were requiring cylinder and boiler renewals. The latter circumstances allowed Thompson to gain the necessary authorisation to fit a similar boiler to the diagram 100 type used by the B17 Class, new cylinders featuring piston valves and long travel valve gear (the settings being worked out in personal conjunction with Thompson and a member of his technical staff), and a new smokebox and cab. The transformed engine was no. 8579 and it entered traffic during May 1932. The engine was proved to be superior to the original B12s, through a 6 lb reduction in coal consumption per engine mile, and 53 more engines were similarly transformed after receiving approval from Gresley, becoming B12/3.

Shortly after the appearance of the locomotive, Thompson returned to the north east and became Mechanical Engineer for the area, with responsibility for Darlington, Stooperdale, Faverdale and Shildon Works falling under him. During his time in this position, Peter Grafton records in *Edward Thompson of the LNER* (2007) that Thompson made an attempt to improve some of the problems in wear and maintenance experienced through the use of Gresley's conjugated motion. In partnership with T.H.W. Cruddas, works manager of Shildon at the

Thompson was the driving force behind the rebuilding of the B12 Class while he was Mechanical Engineer, Stratford. Fifty-four were eventually transformed and no. 8569 became B12/3 in December 1933.

time, Thompson attempted to create the rocking lever through fabrication and welding to reduce the weight of the piece, which was usually forged, and therefore limit the wear on the pin that the arrangement pivoted around. The project was abandoned subsequently as the quality of the welds joining the pieces of the apparatus could not be guaranteed to be satisfactory. It will be noted that this work was carried out with Gresley's permission. Work undertaken at Darlington Works during Thompson's tenure included; rebuilding of the Q5 0-8-0s, D20 4-4-0s and B16 4-6-0s (these latter engines with the Gresley valve gear) and the construction of the V2, D49 and J39 Class locomotives. Thompson would have also been well acquainted with W1

no. 10000, which was often in residence at Darlington Works during this period. In 1938, Thompson made his last move before becoming C.M.E. as he returned to Doncaster to take over from R.A. Thom, who had retired, as Mechanical Engineer, Doncaster. In this position he again attempted to improve the conjugated motion, but these were also unsuccessful as were his plans to rebuild a B3 Class 4-6-0.

When Thompson was appointed C.M.E. the Second World War had been underway for nearly two years and Britain was experiencing its darkest days of the conflict. The railways in general were suffering from arrears of maintenance as there had been a reduction in the number of

Q4 0-8-0 no. 5059 at Doncaster Works in June 1942 to illustrate the class before the rebuilding.

Q1 Class 0-8-0T no. 5058 at Doncaster after being rebuilt from Q4 specifications.

suitable staff to carry this task out due to the requirements of the armed forces. The respective systems were also becoming increasingly important for the transportation of men and materials around the country. Therefore it was vital that as many locomotives as possible were in service and the maintenance required by them was quick and simple to perform for the growing number of inexperienced staff. Consequently, Thompson's aim was to reduce the maintenance requirements of the L.N.E.R's engines to the minimum and improve availability. In the long-term he wanted to significantly reduce the number of classes of locomotives from over 160 to 19 and to standardise these latter as much as possible and make the use of two cylinders

standard practice for engines of low and medium power. At the time of Thompson's appointment A.H. Peppercorn, who had also been with the G.N.R. and L.N.E.R. for a number of years, was made Chief Assistant Mechanical Engineer, while B. Spencer, Gresley's Technical Assistant, was subsequently transferred out of the C.M.E's department.

One of Thompson's first acts to aid maintenance was the removal of the skirting over the coupled wheels of the A4 Class Pacifics. Then, Gateshead works was re-opened to locomotive repairs to improve the availability of the area's engines. The first locomotive to be developed as a standard type was for heavy shunting duties and this was rebuilt from a G.C.R. Robinson Class 8A, L.N.E.R. Q4,

The V4 2-6-2s were Gresley's last design and only two were built before his death. No. 3401 Bantam Cock *and no. 3402 appeared in February and March 1941 respectively.*

0-8-0 engine to become an 0-8-0T locomotive and no. 5058 appeared in June 1942. During the following month the envisaged standard class for freight services was rebuilt from a G.C.R. Robinson 9J Class, L.N.E.R. J11, 0-6-0 engine, no. 6009. Gresley had produced designs for a new three-cylinder 2-6-2 class available for use on large swathes of the L.N.E.R's system just before he died and two of these, no. 3401 *Bantam Cock* and no. 3402, had entered traffic and were to form a numerous V4 Class. Interestingly, the latter had a steel firebox of welded construction with a Nicholson thermic syphon, which were two features also extensively used at this time by Bulleid in his S.R. Pacifics. Thompson discarded Gresley's design and produced his own for the same purpose, incorporating his own design philosophy. The boiler was similar to the diagram 100 used by the B17s, but worked at 225 psi, while the two cylinders were of a standard pattern, as were the 6 ft 2 in. diameter wheels. No. 8301 of the new B1 4-6-0 Class was in service by the end of December 1942.

From Mikados to Pacifics

During their time in service the P2s had encountered some mechanical issues, such as wear of the crank pins of the first pair of coupled wheels and this has been attributed to the pony truck and the double-bolster swing link suspension. This had a tendency to wear rapidly because of difficulty in lubricating the arrangement properly. As a result the suspension's ability to successfully negotiate curves for the coupled wheels was significantly reduced, with the consequence that the leading pair took the brunt

of the forces encountered. Colonel H.C.B. Rogers (1990) records that Chapelon highlighted the pony truck as being a weakness in the design of *Cock o' the North* because he had experienced the locomotive taking curves too roughly while the locomotive was on test in France. An experiment with a K3 Class locomotive, carried out at Doncaster Works not long after *Cock o' the North* was built, highlighted the unpredictable behaviour of the pony truck when under load and, specifically, it was found that the distribution of weight on the pony truck was uneven and favoured one side of the locomotive. Gresley had discarded the double-bolster swing link suspension on the bogies of the Pacifics by the mid-1930s after experimenting with the bogie developed for the D49 Class. Helical spring side control replaced the swing links and the distribution of the locomotive's weight was brought centrally, in addition to larger axle journals and stiffer springs being used. Evidently, Gresley saw no need to extend this modification to the pony trucks despite the problems of wear experienced by a number of classes so fitted. Adding to the difficulties at the front end of the P2 Class was the tendency for the 'big end' of the connecting rod for the middle cylinder to overheat with consequent damage to the crank axle and associated components. The former was also a problem on a number of other Gresley engines, such as the A1, A3 and A4 Class Pacifics. This was in part due to the unequal power outputs of the cylinders which became more pronounced at high speeds, with the centre cylinder working slightly harder, and the design of the big end brasses. The deficiencies in the design of this latter not being satisfactorily remedied until the 1950s. Accusations of track spreading has also been levelled at the class, although no evidence substantiates such a claim.

No. 2003 Lord President *in the midst of a general repair in bay four of the Crimpsall Repair Shop on 3rd April 1938.*

No. 2005 as a Pacific in early 1943.

Plans for the P2s to be rebuilt as Pacifics were either in hand or precipitated by the failure of the crank axle on no. 2004 in April 1942. A number of such failures had occurred up to this date and in the same month the first drawing for the rebuilt locomotive appeared. The pony truck and first pair of coupled wheels were to be removed and replaced by a bogie, with a consequent reduction in the length of the coupled wheelbase to 13 ft and increase of the axle loads of the coupled wheels. The monobloc cylinders were also to be replaced by separate castings and they were staggered so the centre cylinder was well in front of the outside pair. The drive was to be divided so the outside cylinders drove on to the second axle and the centre cylinder to the first coupled axle, with three independent sets of Walschaerts valve gear also being used. Events progressed in the following month as Thompson went to Edinburgh to visit the Scottish Area officials and at this time they were informed that he intended to take the P2s off their current duties. Knox (2011), and several others, inform that a number of attempts were made by E.D. Trask, Locomotive Running Superintendent Edinburgh, and also G. Lund, Technical Assistant Scottish Area, to dissuade Thompson from the rebuild plans, but with no

success. Then, from late May to mid-June, W1 Class no. 10000 was worked between Edinburgh and Aberdeen in an attempt to determine if a powerful Pacific would be suitable for work on the line. When the locomotive had been rebuilt it had been fitted with a diagram 111 boiler which was principally the same as *Wolf of Badenoch*'s boiler, but was working at a higher pressure. The tractive effort of the engine was approximately 7,500 lb more that the A3 and 6,000 lb more than the A4 Pacifics, but with the same adhesion weight of 66 tons. No. 10000's performance was apparently successful, in part due to the fine weather experienced across much of the British Isles during the period, which reduced the possibility of the locomotive slipping.

The necessary authority for the rebuilding of a P2 Class engine was obtained in October 1942. The official reason for the alterations, which was circulated by the L.N.E.R. to a number of the railway journals and magazines subsequently, was that the coupled wheelbase of the class was too long and that it was desirable to reduce this to increase their route availability. No. 2005 *Thane of Fife* was chosen to be the first and the locomotive entered Doncaster Works on the 26th of the month, re-entering

No. 2006 with grey livery applied in April 1944. Note also the locomotive is fitted with a diagram 106A boiler.

No. 2001 Cock o' the North after being rebuilt in September 1944. Note the small smokebox saddle and exhaust ducts from the cylinder to the blastpipe on the outside of the frames. The whistle has been repositioned in front of the cab.

traffic as a Pacific, and nameless, on 18th January 1943. No. 2005 was the prototype for Thompson's envisaged 'A', later A2, Class engines for passenger and freight duties as part of the standard classes to be introduced and was the fourth class to be dealt with out of the ten. It is also pertinent to note that Thompson was to retain the A3 and A4 Pacifics, unaltered from their original design, as well as the V2 Class and V1/V3 2-6-2T locomotives, while the B17s, D49s and K3s would be partially rebuilt with two cylinders as per Thompson's design philosophy. For large three-cylinder locomotives Thompson decided that the Gresley conjugated valve gear was not desirable due to the difficulties in maintaining it satisfactorily in the conditions prevalent, and those for the foreseeable future, and he moved to use three sets of valve gear. To illustrate to the L.N.E.R. board why this was his intention he arranged for Stanier, who delegated the task to E.S. Cox, to compile a report on Gresley's conjugated gear and detailing the problems of wear and the result this had on valve events of the middle cylinder. This advised that when in a complete state of disrepair the centre valve would over travel at high-speed and the cylinder could produce 50% more power than the outside cylinders resulting in stresses over the limit for the big end of the connecting rod. Thompson also provided the board with a dossier of engine failures attributable to this problem. E.S. Cox in *British Railways Standard Steam Locomotives* (1966) notes that in 1941 L.N.E.R. three-cylinder locomotives 'had experienced ten times as many hot bearings on inside as on outside big ends.'

After no. 2005 had been in service for several months, working from at first Doncaster then Haymarket on the usual P2 Class duties, authorisation was given in September 1943 for the conversion of the remaining members of the class. The task began on 28th January 1944 when no. 2006 *Wolf of Badenoch* entered Doncaster Works (the locomotive also re-entered traffic with its nameplates removed, but had them returned, as did no. 2005, in June 1944) and it was then the turn of *Earl Marischal* in April. On 24th June *Cock o' the North* was taken to Doncaster and a start was made on the transformation.

Due to the shortage of resources during the war years as many components as possible had to be retained from the locomotive. Additionally, the necessary consent for new materials had to come from the Ministry of Supply and the Railway Executive Committee and would only be forthcoming for repairs and essential new construction. *Cock o' the North*'s frames were cut between the leading and driving pairs of coupled wheels and the front section was then scrapped. The new front frame section had a length of 18 ft 9¼ in. and was attached to the main frames just rearwards of the first pair of coupled wheels, with the spacing being the same as originally. The overlap of the front and main frames was 4 ft 5¼ in. and the total length of the frames was 44 ft 5¼ in. - 1 foot less than before. The frames at the rear remained the same as previously. New cylinders 20 in. diameter with 10 in. piston valves were produced and the monobloc casting was scrapped. All three cylinders were inclined at 1 in 30 and the drive was divided. The middle cylinder was set well forward, it was 11 ft 8 in. from its driving axle, and an inside motion plate was added and this also provided support for the boiler.

The length of the boiler was shortened by 1 ft 11¾ in. at the front of the parallel section and the distance between the tubeplates became 17 ft, or 16 ft in the case of *Wolf of Badenoch*'s diagram 108 boiler. Incidentally, this was fitted to *Earl Marischal* when the locomotive was rebuilt and the former received the boiler from *Lord President*. *Cock o' the North* received boiler no. 8797 from no. 2002, which had originally been intended for no. 2006 and became the spare boiler when the diagram 108 was introduced. The boiler was reduced in length to improve the layout of the steam passages which would have become complicated if the original dimensions had been retained and they were consequently straightened. The exhaust steam passages proved to be more complex. A route was taken along the exterior of the frames from the outside cylinders before being brought inwards to join with the centre cylinder and then the common chamber for the blastpipe, which was placed inside an unusually long smokebox. The smokebox saddle also had to be altered because of the steam pipe arrangement and was made narrower and longer than the usual Gresley pattern. After modification of the boiler the heating surfaces became; 1,211.57 sq. ft for the small tubes, 1,004.5 sq. ft in the flues and 679.67 sq. ft for the superheater, providing a total heating surface of 3,132

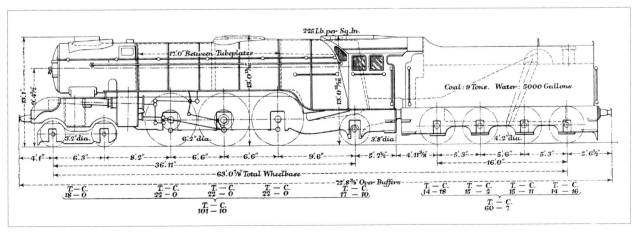

Engine diagram for P2s nos 2003-2006 rebuilt as Pacifics.

sq. ft. For no. 2002's boiler the figures were, respectively; 1,138.7 sq. ft, 995.4 sq. ft, 652.07 sq. ft and 2,988.67 sq. ft. The working pressure of the boiler was increased to 225 psi, while perforated steam collector was necessarily retained. This piece of apparatus was not deemed desirable by Thompson and it was subsequently abandoned by him when possible. The two types of boiler were re-classified diagram 106A and 108A respectively. The Kylchap double blastpipe and chimney were kept, but the dimensions of the arrangement were altered to conform with A3 no. 2751 *Humorist*. The Kylälä spreader was 1 ft 1²⁵⁄₃₂ in. at the bottom and then reduced to 9¹⁄₁₆ in. at the top of the four lobes, while this piece was 1 ft 1³⁄₃₂ in. tall. The cowl above the spreader was 11⁷⁄₃₂ in. diameter and had a height of 1 ft 2⁵⁄₃₂ in. The chimney was 1 ft 4¹³⁄₁₆ in. diameter at the bottom, 1 ft 3 in. at the choke and 1 ft 5³⁄₈ in. diameter at the top. The height from the bottom of the choke was 9¼ in. and from here to the top was 1 ft 7⁷⁄₃₂ in. The chimney was 3 ft 8½ in. tall. Smoke deflectors, in the form of wing plates welded in position, extended 5 ft 6 in. along the top of the smokebox and these were longer by 1 ft 2½ in than the similar arrangement used by no. 2751 at the time.

The coupled wheels, trailing wheels and their springs were retained. The bogie was not of the Gresley D49 pattern, but based on the type developed by Thompson for use on his B1 Class. The load of the engine was passed to the bogie frames at the side through side bearers, with spherical surfaces and bronze slippers, rather than centrally as in the Gresley bogie. The bogie frames were spaced the same distance as the front of the frames and 1 in. thick frame plates were used. Thompson used laminated springs with 7 plates and these were 4½ in. wide by ½ in. thick. The helical horizontal springs had an initial load of 2 tons and the bogie was allowed to move 4 in. either side of the engine. The weight of the engine was reduced to 101 tons 10 cwt and the adhesive weight was 66 tons, while the adhesion factor was 3.67, reduced from the 4.15 of the P2s. Tractive effort was also cut to 40,318 lb. The connecting rods were re-used and three sets of Walschaerts motion were present to operate the valves. As the '2 to 1' lever had been eliminated the maximum cut-off was increased to 75%, while the valve settings for all three cylinders remained as it had been for the piston valve P2s. Also preserved were the cab, tender and regulator. *Cock o' the North* returned to traffic in September 1944 and briefly worked from Haymarket before being reallocated to Aberdeen at the end of October. Following no. 2001 into Doncaster Works was *Mons Meg* and the locomotive had been transformed by November. *Lord President* was in traffic as a 4-6-2 by mid-December 1944.

When all six members of the class had returned to service in Scotland the allocations were *Cock o' the North* and *Earl Marischal* at Aberdeen and nos 2003-2006 at Haymarket. The former engines were based in Aberdeen despite the concentration of the A1, A3 and A4 Pacifics at Haymarket for maintenance reasons after the onset of hostilities. The A2 Pacifics were given maximum loadings

which were the same as the P2s when they re-entered service. However, this was a figure that would have been rarely entrusted to the class and the actual loads generally given to the locomotives were 200 tons less than the limit. During the war years the P2s and then the A2s were aided on the Edinburgh to Aberdeen road by the A1s, A3s and V2s, with the A4s also being recorded travelling to Dundee and even through to Aberdeen. Through working had been re-introduced during this time and footplatemen were generally swapped at Dundee. When the war ended the duties of the class were revised and altered so that the principal tasks for the locomotives were goods, parcels, fish and meat trains, with the number of appearances on any passenger service being quite low. The engine's sphere of activity was also transformed and they could now be seen working into England, generally as far south as Newcastle. The diversification of the A2's activities also brought them into contact with different crews from other sheds in the Edinburgh area, Dundee, Aberdeen and the North Eastern Area.

Following the A2s' re-entry into service a number of problems were experienced at the front end of the locomotives and these included breakages involving the steam pipes, exhaust passages, cylinder bolts and smokebox saddle bolts. This cause of these failures was attributed to the front end of the frames being provided with too little support and the curves in Scotland producing too much strain on them. Talking to Colonel H.C.B. Rogers for his book *Thompson & Peppercorn: Locomotive Engineers* (1979) J.F. Harrison, Mechanical Engineer Scotland 1945-1947, notes that there was a 'serious frame weakness' present between the cylinders and leading coupled wheels. While in the same publication T.C.B. Miller, who held a number of posts in Scotland and then England for the L.N.E.R., comments that in service the frames flexed 'up to about 3 inches.' No. 2005 suffered particularly towards the end of the war. An interesting point to note is that over 14 months elapsed following the engine's reconstruction before the first visit to Cowlairs Works occurred in early March 1944 for light attention. Then, two further entries to the repair shop were recorded. The first took up eighteen days between the end of March and early April, while the second lasted for six days in June. *Thane of Fife* underwent its first heavy repair at Cowlairs between 8th August and 14 September 1944. Six months subsequently passed before the first of nine entries into the works for attention and one visit was also made to Inverurie, excluding another one for weighing purposes. A total of 67 days were lost to the engine receiving attention and this does not include time out of traffic on shed or undergoing repairs that did not necessitate travelling to the works. *Wolf of Badenoch* was out of service a similar number of times in 1944 and 1945 and lost 42 and 16 days in works respectively. The remaining locomotives also saw the Scottish Area's repair facility a number of times during 1945, with *Lord President* achieving the highest number at 34 days spread over 5 visits; *Earl Marischal* lost 26 days over 4 and *Mons Meg* 16 days over 3 trips to Cowlairs. As a result of the frame movement a support plate was added to the inside of the frames between the centre and outside cylinders and this was 1 in. thick. At the same time the exhaust passages were divided into two sections and connected by means of an expansion joint, while the live steam pipes were also made longer. These modifications did alleviate the situation somewhat, but shed staff subsequently had to remain vigilant for any repeats of the problems.

Cock o' the North was only in Cowlairs Works for light repairs once during 1945 and this lasted for 15 days between 22nd September and 6th October. It was a further six months before no. 2001 entered Doncaster for its first general repair as an A2/2 (the locomotives' new classification from mid-1945) and this took place between 30th March and 4th May 1946. Then, at the end of the month, *Cock o' the North* entered Cowlairs to be the first of the class to receive the L.N.E.R. apple green livery with lining after the end of the war. This task was performed so that the engine could take part in an exhibition of rolling stock in Edinburgh, this being similar in nature to those taking place prior to the war, but this was to be the last appearance of the locomotive at such an event.

A2/1s and Testing

After the emergence of *Thane of Fife* in 1943, Thompson set plans in motion to replace the pony truck of the final four V2 Class engine on order with a bogie so they became Pacifics. By the end of the year the required authorisation had been granted by the L.N.E.R. board and the work put in hand at Darlington Works. Again, as much of the material originally to be used on the engines was retained, but some alterations were made. The boiler was of the V2 diagram 109 design with the only change being to the working pressure which was increased by 5 psi to 225 psi.

The grate area also remained the same, but an addition was made through the provision of a rocking grate and hopper ashpan for ease of maintenance. The Kylchap double chimney and blastpipe was added, being of the dimensions of the previous applications and the engines initially had the wing type deflectors as fitted to the A2/2s, although these would later be altered to the larger type. The cylinders were cast separately and also staggered as in the A2/2 design. The diameter of the cylinders was reduced from the latter class, but increased from the 18½ in. of the V2s to 19 in. diameter with 10 in. piston valves, which were increased in size by 1 in. from the standard V2 dimension. The conjugated motion was dispensed with as three sets of Walschaerts valve gear were provided (the maximum cut-off was again increased to 75%) and steam reversing gear was introduced, replacing the Gresley screw-type reverser. The bogie was similar to the B1 type and the cabs were v-shaped, but some of the interior details were changed. The Gresley two-tier floor and regulator handle were replaced by a level floor and a G.C.R. type regulator handle. A new feature applied to three of the four locomotives was electric lighting powered by an axle generator for the cab interior and to replace the headlamps (there were four white lights and a red one) for displaying the headcodes at night. The weight of the engines was increased to 98 tons, whereas the V2 weighed 93 tons 2 cwt. The tractive effort also increased from 33,730 lb to 36,387 lb. The engines were constructed between May 1944 and January 1945, taking the numbers 3696-3699 and later being bestowed with names with a Scottish origin; *Highland Chieftain*, *Duke of Rothesay*, *Waverley* and *Robert the Bruce* respectively. When introduced no. 3696 was classified 'A' as the P2 rebuilds had been, but in mid-1944 they became A2/1. The class also had to have the front of the frames strengthened as they were undergoing their first general repairs.

A2/1 no. 3696 originally had a six-wheel tender, but later acquired tender no. 5672 from A4 no. 4469 Sir Ralph Wedgwood. *The pairing has been recorded at Barkston on 20th December 1945.*

With two Thompson Pacific classes in service, the decision was taken to perform trials for comparative purposes. They were scheduled to take place in early 1945 and the locomotives selected were no. 2003, no. 3697 and A4 no. 2512 *Silver Fox*. The A2/2 and A4 were tested with the King's Cross to Leeds passenger service and the return on the following day, the dates being the 15th and 16th and 22nd and 23rd January respectively, while the A2/1 took part on 29th and 30th. The loads were heavy in all instances. *Lord President* had 19 vehicles, 596 tons, and 18 carriages, 578 tons on the two days, although these were considerably reduced at Doncaster, this being true for all the engines. *Silver Fox* had similar train weights on both days and no. 3697 was coupled to services loaded to 579 and 548 tons. On both no. 2512 and no. 3697's return journeys to London mechanical issues, in the form of water scoop damage and a broken steam heating pipe respectively, brought the participation of the engines on this trial to an end. Therefore, only the coal consumption figures produced on the journey to Leeds are relevant and these were; no. 2003, 70.5 lb per mile, no. 2512, 52.3 lb per mile and 69 lb per mile from no. 3697. However, adverse weather and running conditions occurred on these journeys and must be taken into account. No. 2003 and no. 3697 were also able to run the scheduled freight service from King's Cross to Doncaster and this was very heavily loaded to between 54 and 60 wagons weighing from 771 tons to 860 tons. The average coal consumption for no. 2003 was 68.7 lb per mile and no. 3697's figure was 63.8 lb. In addition to the above, the A2/2 and the A2/1 were able to complete the scheduled additional passenger service trials between King's Cross

and Grantham. The trains on these trips weighed 487 and 604 tons and 507 and 578 tons respectively. The coal consumption figures on these occasions averaged 62.8 lb for *Lord President* and 60.2 lb from no. 3697. Due to the problems experienced returning from Leeds the tests, not being considered truly representative of the engines' optimum performance, were abandoned and they were rescheduled for later in the year.

The tests subsequently took place at the end of April and lasted to the beginning of June with the three engines again participating. The services were the 10.30 a.m. King's Cross to Grantham and return at 3.17 p.m.; 12.25 a.m. King's Cross to Peterborough express goods, returning at 4.50 a.m. and this schedule was worked for a week with one crew attached to each engine. All three locomotives were heavy on coal at the beginning of the week, but by the end it had been reduced. No. 2512 produced the best figures for the passenger service trains, which were predominantly formed by 18 carriages weighing between 500 and 600 tons, with a coal consumption of 38 lb per mile. No. 3697 was second with 41.69 lb per mile and last was no. 2003 with 42.73 lb. The overall average for the week of working the service was, for each locomotive per engine mile respectively; 40.5 lb, 43.3 lb and 46.1 lb. No. 3697 was considerably more economical on the freight services, which consisted of trains 50-60 wagons long and between 500 and 760 tons in weight, with a best figure of 36.75 lb per mile and next was *Lord President* with 40.89 lb. *Silver Fox* consumed 45.7 lb per mile. The average figures brought the A2/2 and A4 closer together - 45.2 and 45.4 lb per engine mile respectively - while the A2/1 was still in front with a coal consumption of 41.4 lb per mile.

A4 no. 2512 Silver Fox *in Doncaster Works' Paint Shop. The locomotive was tested against no. 2003 and 3697 during the first half of 1945.*

No. 500, the first new Thompson Pacific, outside the Paint Shop.

A2/3s and Other Locomotives

New locomotives to the A2 design were authorised in mid-1944 and again towards the end of the following year bringing the total to be built to 43. In the event, however, only 15 engines were constructed and they were erected by Doncaster Works between May 1946 and September 1947. Numbered 500, 511-524, the engines were generally similar to the A2/2s, but with a few minor alterations. The boiler pressure was increased to 250 psi and there was a small increase in the length of the firebox combustion chamber altering the firebox heating surface to 245.3 sq. ft. Also, the perforated steam collector was replaced by a conventional steam dome and these changes led to the new boiler being classified diagram 117. A modification was carried out to the Kylchap blastpipe tops where the vee bars were cast integrally with the top as a means to reduce maintenance of the apparatus. Large smoke deflectors were also introduced as the wing type were proving to be ineffective. As a further aid to reducing the attention required by the locomotive, wire screens were placed at an angle in the smokebox to help break up ash and other waste so it could be ejected from the chimney rather than accumulate in the smokebox. The diameter of

A2/3 no. 512 Steady Aim *in August 1946. Note the steam dome and five boiler wrapper plates.*

Gresley's pioneer Pacific, no. 4470 Great Northern, is pictured in September 1945 in its original rebuilt condition with short cab sides, no smoke deflectors, electric lighting (driven from the rear bogie axle) and double chimney with beading. No. 4470's livery was also a departure from L.N.E.R. practice as royal blue with red lining was chosen.

Thompson Class D two-cylinder 4-4-0 no. 365 The Morpeth, B.R. no. 62768, is photographed at Harrogate station on 29th November 1951 and would be withdrawn a year later following a crash. Picture courtesy of Yorkshire Post.

the cylinders was reduced to 19 in., but the piston valves remained at 10 in. diameter and the tractive effort was slightly increased to 40,430 lb. A cab with a flat front was given to the engines in place of the v-shaped cab. No. 500 was the first to be completed, with Doncaster Works no.

2000, and was bestowed with the name *Edward Thompson* at a naming ceremony at Marylebone station on 31st May 1946. Thompson was afforded this honour as he was about to retire at this time and to recognise his services to the L.N.E.R. throughout his career.

No. 8301 Springbok *was the first of Thompson's B1 Class 4-6-0s and was built at Darlington in December 1942.*

B2 no. 1671 Royal Sovereign.

Upon Thompson leaving office his standardisation plans were progressing steadily as members of several classes were under reconstruction to form his standard fleet of engines, in addition to new locomotives being built to contribute to their numbers. With regards the former engines, a substantial percentage were being rebuilt with the standard boiler of the B1 Class and their two cylinders, which were 20 in. diameter with 10 in. piston valves. A number of the rebuilds were prototypes and served evaluation purposes before features were incorporated into the standard design, therefore only one engine was rebuilt. Such examples included the B3/2 to B3/3 conversion, D49/2 to Class D 4-4-0 and K3 to K5. In addition, the first A1 Class Pacific, no. 4470 *Great Northern*, was transformed as the first of a projected new express passenger class, which was an extremely controversial decision to say the least. No. 4470 had the same front end layout as the A2s, but the boiler was the same as used by the A4s and a Kylchap blastpipe and chimney arrangement was also fitted.

The more numerically substantial classes were the O1s, which had been rebuilt from the copious number of O4s possessed by the L.N.E.R., and eventually had 58 members by 1949 when the programme was discontinued. Seventeen B16 Class engines were reconstructed to become B16/3 and ten B17s were altered and consequently reclassified B2. These engines had the Thompson B1 features in addition to his type of bogie. A Gresley K4 Class engine (introduced in the late 1930s for the West Highland line which was similar in nature to the Edinburgh to Aberdeen line) was also rebuilt to provide an initial template for the new mixed traffic engine to replace the K4s, J38 and J39 Classes. This locomotive was the sole example, but it was used for the basis on the K1 Class subsequently built and numbered 70 engines. The new classes were; the B1 4-6-0s, which subsequently reached a total of 410, the A2 Pacifics and L1 2-6-4T for mixed traffic purposes, later amounting to 100. However, Thompson's plans would be unfulfilled because his successor was more inclined towards Gresley's design ideas and in the near future there were changes to the rail industry in the form of Nationalisation. The new organisation would have its own views on rolling stock and standardisation, although their view of maintenance was quite similar to that of Thompson.

Thompson L1 2-6-4T no. 9000.

Peppercorn and the A2/2s in B.R. Service

Thompson was succeeded in the C.M.E. position by A.H. Peppercorn, who had the design for the remaining A2s on order revised. Fifteen new engines were constructed and they saw the return of a number of features. The perforated steam collector was reinstated to the top of the boiler, but it retained the particulars of the diagram 117 boiler and a reclassification to diagram 118 ensued. Also returning were the v-shaped cab front and the cylinders to the conventional position between the bogie wheels, with a consequent reduction in the size of the smokebox. The Thompson divided drive was retained as were the three sets of valve gear. The steam passages were redesigned as a result of the front end movements and to reduce problems of steam leakage experienced by the Thompson Pacifics. The Kylchap double blastpipe and chimney was abandoned initially on all but the last member of the class, no. 60539 *Bronzino*. However, five more members of the class subsequently acquired the apparatus, in addition to an M.L.S. multiple valve regulator. The engines were constructed at Doncaster between December 1947 and August 1948 and classified A2 with the Thompson A2s made A2/3.

With 30 new Pacifics in service, and after the dust had settled from Nationalisation of the railways from 1 January 1948, a movement of engines occurred. The A2/2s at Haymarket, which had been joined by *Cock o' the North* and *Earl Marischal* from September 1949, were moved to England at the end of 1949, to the beginning of 1950. No. 2001, no. 501 from August 1946 and British Railways no. 60501 from May 1948, was sent to York on 27th November 1949, with no. 2002, which had become

Arthur Henry Peppercorn (1889-1951) C.M.E. of the L.N.E.R., 1946-1948.

Peppercorn A2 Class Pacific no. 525 leaves Doncaster Works early in December 1947 shortly after being completed for its first trial run.

By 31st December 1947 no. 525 had acquired the name A.H. Peppercorn in honour of the designer and last C.M.E. of the L.N.E.R. Peppercorn is in the centre right and on his left is J.F. Harrison (a pupil of Gresley's), Assistant Chief Mechanical Engineer, who later became C.M.E. of B.R. in 1958. Interestingly, prior to this appointment he was part of the team headed by R.A. Riddles that developed B.R. Standard Class 8 Pacific no. 71000 Duke of Gloucester, which was fitted with British Caprotti poppet valves and rotary cam valve gear. Gresley's long-serving Technical Assistant Bert Spencer is second from the right on the front row.

No. 60505 Thane of Fife *awaits entry to the Crimpsall Repair Shop on 26th April 1950 for a general repair, its first since the same month of 1948 and the locomotive's first at Doncaster since 1946.*

no. 60502, and *Lord President*, no. 60503. No. 2004 (60504) was reallocated to Peterborough New England shed in January 1950 to join no. 60505 *Thane of Fife* and no. 60506 *Wolf of Badenoch*, which had been at the shed since November 1949. Five Peppercorn A2s replaced them at Haymarket, joining five other members of the class in

the country since mid-1949, as well as Thompson A2/1 no. 3696, B.R. no. 60507 *Highland Chieftain*. Thompson A2/3 no. 519, B.R. no. 60519 *Honeyway* went new to Haymarket in February 1947 and also spent many years working from the depot. The Peppercorn A2s were based not only at Haymarket, but Dundee and Aberdeen and

No. 60529 Pearl Diver *was one of six Peppercorn A2s recruited to Haymarket between February 1948 and January 1951 to replace the A2/2s. No. 60529 is pictured at the depot on 2nd July 1961. Photograph courtesy of Bill Reed.*

A2/2 no. 60501 Cock o' the North poses for its last official picture after completing a general repair at Doncaster Works in February 1950. No. 60501 had previously acquired its smokebox numberplate in April 1949, resulting in the raising of the top lamp iron, but has now lost the metal plates around the sand fillers and this would seem to have remained the arrangement until withdrawal. Cock o' the North had the honour of being the first of the class to have B.R's Brunswick green livery applied with black and orange lining; the last was no. 60502 in March 1951.

No. 60501 Cock o' the North *passes Arksey, near Doncaster, on 20th June 1952 with a Colchester to York express service. Picture courtesy of* Yorkshire Post.

worked much of the express passenger traffic on the Edinburgh to Aberdeen line, which was now averaging approximately 450 tons in weight rather than the heavy loads the P2s were expected to handle, as well as services on radiating lines to other population centres. The A2/3 also had a varied life at Haymarket, working both passenger and goods trains, as did the three A2/1s; no. 3698, later B.R. no. 60509 *Waverley* and no. 3699, B.R. no. 60510 *Robert the Bruce* were allocated to Haymarket from March 1945 and March 1949 respectively.

An event worth noting here is the opening of Rugby testing station, which occurred several months earlier in October 1948, and, fittingly, L.N.E.R. A4 Class no. 4498 (B.R. no. 60007) *Sir Nigel Gresley* was the first locomotive to ride the rollers. The L.N.E.R. and L.M.S. had entered into an agreement in 1937 to construct a locomotive test plant at Rugby after an intention to use a site near Leeds was abandoned. The design for the plant was based on the facility at Vitry and before the outbreak of war, Heenan & Froude Ltd were tasked with supplying the brakes

With some steam leaking from the front end, Cock o' the North *disturbs the quiet of Arthington station, near Harrogate. The smokebox number plate has been lowered to the top door hinge to allow the top lamp iron to take up its original position. This modification dates from September 1956. Picture courtesy of* Yorkshire Post.

A4 Pacific no. 60007 Sir Nigel Gresley *positioned on the rollers at Rugby testing station for the opening in October 1948.*

No. 60502 Earl Marischal *at Grantham shed on 8th May 1960. The locomotive has a diagram 118 boiler fitted with five wrapper plates, lipped chimney, smokebox number plate in the lower position and no back plates for the sand fillers. Picture courtesy of Bill Reed.*

No. 60506 Wolf of Badenoch *was the only A2/2 not to have the chimney swapped for the lipped variety. No. 60506 is at Doncaster shed during September 1958, possibly before or after a visit to the works for light repairs. Picture courtesy of Bill Reed.*

No. 60504 Mons Meg *rests at Bridge Junction, Doncaster on 25th May 1958. Picture courtesy of Bill Reed.*

Bridge Junction, Doncaster, looking north towards the station, on 16th March 1951 after the derailment of the 9.13 a.m. York to King's Cross service hauled by no. 60501 Cock o' the North.

and Amsler commissioned to supply the dynamometer. Construction was understandably suspended during hostilities, but an effort was made to resume work soon after cessation of the conflict. When completed the testing plant had cost approximately £150,000. Only two L.N.E.R. locomotives were thoroughly tested at Rugby, D49/2 62764 *The Garth*, which was fitted with Reidinger valve

gear to operate the poppet valves at an infinitely variable number of cut-offs, and Thompson B1 no. 61353. A number of L.M.S. engines underwent trials, including Hughes 'Crab' Class engine fitted with poppet valves and the Reidinger gear and a Stanier Class Five 4-6-0 equipped with Caprotti valves. The B.R. Standard Classes were also extensively tested.

The third and fourth carriages were the first to leave the track and, unfortunately, this pier supporting Balby Road Bridge was in their path and caused significant damage to both as is evident here.

At York the A2/2s could be put on passenger services to the capital, while other duties included passenger services to Newcastle and to Leeds. The goods diagrams were similar, including parcels and mail trains, but extended to Grantham, Peterborough and Darlington. *Locomotives of the L.N.E.R. Part 2A* (1973) records that *Cock o' the North* managed to extended its working range even further shortly after arriving at York as it was noted at Nottingham Victoria station with a long train of empty coaches in 1950 and then shortly afterwards the engine was again on the G.C. line heading a football special to Wembley. No. 60501 and no. 60503 were also allocated away from York at Leeds Neville Hill depot for a time as two of the shed's Gresley Pacifics were out of action and this loan lasted from 27th November to 17th December 1950. The locomotives at Peterborough could work as far afield as Newcastle and York with both passenger and freight trains, while the passenger services to and from London could be both express and stopping trains.

While working from York on 16th March 1951 *Cock o' the North* was involved in a serious derailment at Bridge Junction, a short distance away from Doncaster station. The engine had brought the 9.13 a.m. York to King's Cross service, which consisted of eight carriages and a horsebox, to Doncaster where the 8.45 a.m. from Hull, made up of six carriages, was attached to the front of the train from York. The combined weight of the trains was 448 tons tare and 470 tons gross and there were an estimated 250 passengers on board. The train left Doncaster station's platform four one minute late from the 10.06 a.m.

departure time and went via the slow line to join the 'up' main line just north of Bridge Junction. As this switching of lines occurred the third carriage of the train derailed at the rear end of the scissors crossover and the next six carriages were subsequently taken off the track in the same area. The derailed carriages were then pulled forward by the locomotive for 60 yards until the track formation forced the third carriage on to its side before colliding with a pier supporting Balby Road bridge, causing significant damage to the carriage and to the fourth carriage. The coupling between the second and third carriages then snapped and the locomotive was brought to a halt 75 yards further along the line after the crew were alerted to the accident. Some of the other derailed carriages remained upright while others were also forced on to their sides. Sadly, 14 passengers were killed, while 12 were seriously injured and 17 others sustained minor injuries.

Cock o' the North was cleared of any blame in the derailment when the ensuing report was issued. The engine had run 3,000 miles since its last heavy intermediate overhaul at Doncaster, which took place from 5th January to 16th February, and no faults were found prior to the engine leaving York shed on the morning in question or when inspected after the accident. The carriages were also similarly in a generally sound condition. A speed limit of 10 m.p.h. was in force on the slow line and through the scissors crossover because the cant gradient on these two lines was at odds with the main line, which was steeper. The report highlighted the speed the train was travelling as contributing to the severity of the accident. The signalmen

With the derailment being close to Doncaster Works and station the response to the accident was immediate and help arrived quickly for the injured.

and eyewitnesses estimated the speed to have been between 10 and 25 m.p.h. through the crossover, while the driver of the engine straight after the derailment said that the speed could have been near 25 m.p.h. However, he later reversed this and added that he had passed through the crossover at up to 20 m.p.h. on previous occasions. A number of other engines were noted to have passed through without maintaining the limit. Tests were subsequently carried out to determine the speed, these involving *Cock o' the North,* which had not been fitted with a speed indicator at the time of the accident, and other Pacific engines. Through these the speed was established to have been near 15 m.p.h., but could have been as high as 20 m.p.h. The report noted that while the speed was excessive with regards the restriction there was no way that the outcome could have been predicted and the derailment would have been much less severe had the bridge not been so close at hand.

The reduction of the cant gradient was necessary on the transition between the two lines through the crossover by using 1 in. thick oak packing pieces under the chairs of the left-hand rails. One of these pieces was decayed and split only a short distance from the check rail. The crossover was about a week away from undergoing its monthly thorough repair and had been inspected the day before the accident by the Acting Ganger responsible for it and only a few days previously by the Permanent Way Inspector for the Doncaster Area. At the beginning of the trailing-v crossing one of several bolts holding the crossing together was found to be missing and on the opposite side another bolt was

loose. Two bolts were broken during the derailment and after the other bolts from the crossing were examined they were found to have fatigue fractures. A number of sleepers were also found to be 'pumping' near the crossover. The failure of the right-hand wing rail in the crossing was deemed to have been the cause of the derailment due to its weakening through fatigue and the missing bolt. The failure of the bolts was deemed difficult for the permanent way staff to detect, but the report said that they should have been paying closer attention to the crossover given that the condition of some of the packing pieces was poor, the sleepers were pumping and a bolt was missing.

As with the P2s, the times of the A2/2s were seldom recorded in the various publications. One of the few journeys noted in any detail concerns *Earl Marischal* and its journey with a portion of the King's Cross to Newcastle service. The engine came on to the train at Doncaster and carried it through to Darlington non-stop. The timings were reproduced in C.J. Allen's *British Pacific Locomotives* (1975) and made between York and Darlington, with the return service also given, illustrating the locomotive's capacity for high-speed running. The load was light, 220 tons total, and no. 60502 took the train through York at 20 m.p.h., then ran a mile-a-minute for the following 11-mile level section to Alne, where the speed had reached 78 m.p.h. Just over 10 miles later at Thirsk the engine was 30 seconds under the 20 minute schedule for the section and was travelling at 84 m.p.h., having maintained figures of above 80 m.p.h. for the length of the level stretch of line between Pilmoor and Thirsk. Northallerton, 30 miles from York, was passed in 24 minutes 56 seconds, a saving of 1 minute 4 seconds on the prescribed time, and at 76 m.p.h. The following 14 miles to Darlington were similarly travelled at high speed despite there being a few more undulations in the gradient profile than previously and the final time saw a 3 minute 17 second reduction of the 40 minutes 30 seconds booked. A similar performance was garnered on the return journey. Despite losing nearly 2 minutes between Darlington and Eryholme, where there was a section of track rising for approximately two miles at 1 in 391 to the later place, which no. 60502 passed at 56 m.p.h., time was progressively reduced to just over 1 minute when passing Northallerton at 77 m.p.h. The deficit was just 30 seconds 7 miles later as *Earl Marischal* went through Thirsk at 92 m.p.h. Speeds above 90 m.p.h. were maintained on the level plains to Alne and by this point time favoured the engine by 22 seconds. The remaining 11 miles to York took 8 minutes 47 seconds to complete as no. 60502 passed through the station at 30 m.p.h. and 1 minute 35 seconds in front of the schedule.

By the early 1950s the diagram 106A boilers (the diagram 108A boiler had been scrapped in mid-1946) were in need of extensive repairs and because of the loss of the above there was no spare boiler. This contributed to the class having lengthy spells out of traffic in the late 1940s and into the following decade, with the result being that the fitting of diagram 118 boilers, as used on the Peppercorn Pacifics, and similarly for the A2/3s, was sanctioned. *Cock o' the North* carried a number of diagram 118 boilers after its boiler from rebuilding was removed in October 1952, and this was next used by *Mons Meg*. Incidentally, no. 60501's original boiler, renumbered 29770, was paired with no. 60504 until its replacement in this instance and afterwards *Lord President* received boiler no. 29770. Generally speaking, the A2/2s required more works visits than the other Thompson Pacific classes during the 1950s as both the A2/1s and A2/3s averaged about 80,000 miles between general repairs. For example *Cock o' the North* was

in Doncaster Works five times between November 1953 and May 1954 for non-classified repairs, before undergoing a general repair between August and September and the engine was again in works briefly before the end of the year. Although, no. 60501 then proceeded to be in service for 20 months before returning to Doncaster for its next general overhaul, accumulating 91,500 miles in traffic, this appears to be have been an exception. York A2/3 no. 60512 *Steady Aim* underwent just one general repair during the same period as did no. 60524 *Herringbone*. Peterborough A2/3 no. 60514 *Chamossaire* also held a similar record against *Mons Meg* for the period 1955-1957 as did A2/1 no. 60508.

In the aftermath of B.R's Modernisation Plan of 1955 steam locomotive classes consisting of a small number of engines were marked for withdrawal when heavy repairs were due in order to make way for a new generation of diesel locomotives. A start had been made on the A2/2 Class in November 1959 when *Thane of Fife* entered Doncaster Works and was condemned, followed by *Lord President*, which had run the lowest mileage of the class with 508,500 miles, in addition to having the second lowest mileage for when running as a P2. *Cock o' the North* was the next engine to be withdrawn in February 1960 and had run 616,461 miles since being rebuilt or a total of 978,597 miles. The remaining members of the class were removed from traffic in 1961; *Mons Meg* in January, *Wolf of Badenoch* in April and *Earl Marischal* in July. No. 60502 had accomplished just over 1 million miles in service. The A2/1s were taken out of service between August 1960 and February 1961 having acquired slightly higher mileage figures than the A2/2s. The lowest was no. 60508 with 749,952 and the highest was no. 60509 with 818,943. In November 1962 no. 60113 *Great Northern* was withdrawn and just fell short of the million-mile mark since being rebuilt, while in the same month the first A2/3 was sent to the scrapyard; no. 60519 *Honeyway* was condemned before the end of the year and had run over 900,000 miles in Scotland. The Thompson Pacifics became extinct in June 1965.

The end of the line for no. 60501 Cock o' the North. After being condemned in February 1960 the locomotive stands on the scrap line at Doncaster Works with L.M.S. Fowler 2P Class 4-4-0 no. 40582 on 20th March 1960. All of the A2/2s were broken up at Doncaster.

Bibliography

Allen, Cecil J. *British Pacific locomotives*. 1975.

Bellwood, John and David Jenkinson. *Gresley and Stanier*. 1986.

Bonavia, Michael R. *A History of the LNER: 2. The Age of the Streamliners, 1934-39*. 1985.

Brown, F.A.S. *Nigel Gresley: Locomotive Engineer*. 1962.

Bulleid, H.A.V. *Bulleid of the Southern*. 1977.

Clay, J.F., Ed. *Essays in Steam*. 1970.

Clay, J.F. and J. Cliffe. *The LNER 2-8-2 and 2-6-2 Classes*. 1973.

Coster, Peter. *The Book of the A1 and A2 Pacifics*. 2007.

Cox, E.S. *British Railways Standard Steam Locomotives*. 1966.

Grafton, Peter. *Edward Thompson of the LNER*. 2007.

Haigh, Alan J. *Locomotive Boilers*. 2003.

Harris, Michael. *Gresley's Coaches: Coaches Built for the GNR, ECJS and LNER 1905-53*. 1973.

Harris, Michael. *LNER Carriages*. 2011.

Hughes, Geoffrey. *A Gresley Anthology*. 1994.

Hughes, Geoffrey. *The Gresley Influence*. 1983.

Hughes, Geoffrey. *LNER*. 1987.

Hughes, Geoffrey. *Sir Nigel Gresley: The Engineer and His Family*. 2001.

Knox, Harry. *Haymarket Motive Power Depot, Edinburgh: A History of the Depot, its Works and Locomotives 1842-2010*. 2011.

Marquess of Huntly. *The Cock o' the North*. 1935.

Marshall, Peter. *The Railways of Dundee*. 1996.

Nock, O.S. *British Locomotives of the 20th Century: Volume 2 1930-1960*. 1984.

Nock, O.S. *The Locomotives of Sir Nigel Gresley*. 1945.

Pike, S.N. *Mile by Mile on the L.N.E.R. King's Cross Edition*. 1951.

RCTS. *Locomotives of the LNER: Part 1 Preliminary Survey*. 1963.

RCTS. *Locomotives of the LNER: Part 2A Tender Engines – Classes A1 to A10*. 1978.

RCTS. *Locomotives of the LNER: Part 2B Tender Engines – Classes B1 to B19*. 1975.

RCTS. *Locomotives of the LNER: Part 3A Tender Engines – Classes C1 to C11*. 1979.

RCTS. *Locomotives of the LNER: Part 3B Tender Engines – Classes D1 to 12*. 1980.

RCTS. *Locomotives of the LNER: Part 3C Tender Engines – Classes D13 to D24*. 1981.

RCTS. *Locomotives of the LNER: Part 4 Tender Engines – Classes D25 to E7*. 1968.

RCTS. *Locomotives of the LNER: Part 5 Tender Engines – Classes J1 to J37*. 1984.

RCTS. *Locomotives of the LNER: Part 6A Tender Engines – Classes J38 to K5*. 1982.

RCTS. *Locomotives of the LNER: Part 6B Tender Engines – Classes O1 to P2*. 1991.

RCTS. *Locomotives of the LNER: Part 6C Tender Engines – Classes Q1 to Y10*. 1984.

RCTS. *Locomotives of the LNER: Part 7 Tank Engines – Classes A5 to H2*. 1991.

RCTS. *Locomotives of the LNER: Part 8A Tank Engines – Classes J50 to J70*. 1970.

RCTS. *Locomotives of the LNER: Part 8B Tank Engines – Class J71 to J94*. 1971.

RCTS. *Locomotives of the LNER: Part 9A Tank Engines – Classes L1 to N19*. 1977.

RCTS. *Locomotives of the LNER: Part 9B Tank Engines – Classes Q1 to Z5*. 1977.

Rogers, Colonel H.C.B. *Express Steam Locomotive Development in Great Britain and France*. 1990.

Rogers, Colonel H.C.B. *Thompson & Peppercorn: Locomotive Engineers*. 1979.

Smith, Martin. *The Gresley Legacy: A Celebration of Innovation*. 1992.

Yeadon, W.B. *Yeadon's Register of LNER Locomotives Volume One: Gresley A1 and A3 Classes*. 2001.

Yeadon, W.B. *Yeadon's Register of LNER Locomotives Volume Two: Gresley A4 and W1 Classes*. 2001.

Yeadon, W.B. *Yeadon's Register of LNER Locomotives Volume Three: Raven, Thompson & Peppercorn Pacifics*. 2001.

Yeadon, W.B. *Yeadon's Register of LNER Locomotives Volume Nine: Gresley 8-Coupled Engines Classes O1, O2, P1, P2 & U1*. 1995.

Young, John and David Tyreman. *The Hughes and Stanier 2-6-0s*. 2009.

A Railway Engineer. 'Some Footplate Runs on the L.N.E.R. 2. - North British "Atlantics."' *The Meccano Magazine*. June 1935 pp. 346-347.

A Railway Engineer. 'Footplate Runs on the L.N.E.R. V. - The "Mikados". *The Meccano Magazine*. October 1935 pp. 586-588.

'A.C.F.I. Feed-water Heating Apparatus, L.M.&S. Ry.' *The Locomotive Magazine & Railway Carriage & Wagon Review.* Volume 38, October 1932, p. 359.

Allen, C.J. 'Britain's First Eight-Coupled Express Engine - L.N.E.R. 2-8-2 Locomotive No. 2001, "Cock o' the North."' *The Railway Magazine.* Volume 75, No. 445, July 1934, pp. 33-37.

Allen, C.J. 'British Locomotive Practice and Performance.' *The Railway Magazine.* January 1935, pp. 12-23.

Allen, C.J. 'British Locomotive Practice and Performance.' *The Railway Magazine.* February 1935, pp. 87-95.

Allen, C.J. 'British Locomotive Practice and Performance.' *The Railway Magazine.* January 1936, pp. 13-23.

'An Improved Locomotive Feed-Water Heater and Pump.' *The Locomotive Magazine & Railway Carriage & Wagon Review.* Volume 34, July 1928, pp. 225-226.

'An L.N.E.R. Cock o' the North Class Locomotive Rebuilt as a Pacific.' *The Railway Magazine.* Volume 90, 1944.

'At Harwich - Huge Locomotive.' *Harwich and Manningtree Standard.*

Bulleid, O.[V.S.] 'Poppet Valves on Locomotives.' *Journal of the Institution of Locomotive Engineers.* Volume 19, Journal No. 90, Paper No. 248, 1929, pp. 569-623.

Bulleid, O.V.S. 'Locomotives I Have Known.' *Proceedings of the Institution of Mechanical Engineers.* Volume 152, pp. 341-352.

Caprotti, A. 'A New Locomotive Distributing Gear Using Poppet Valves.' *Journal of the Institution of Locomotive Engineers.* Volume 15, Journal No. 68, Paper No. 176, 1924, pp. 86-119.

Churchward, G.J. 'Testing Plant on the Great Western Railway at Swindon'. *Proceedings of the Institution of Mechanical Engineers.* Volume 67, June 1904, pp. 937-939.

'"Cock o' the North" - Secrets of One of Britain's Most Famous Locomotives.' *Railway Wonders of the World.* Part 13, April 1935, pp. 400-405.

'"Cock o' the North."' *The Railway Engineer.* Volume 55, August 1934, pp. 233-243.

'Cock o' the North Class Locomotive Rebuilt as a Pacific.' *The Railway Gazette.* Volume 80, March 1944, pp. 311-312.

Coggon, A.P. 'London & North-Eastern Railway: The Class "A.2" Conversion From "P.2."' *Journal of the Stephenson Locomotive Society.* July 1943, p. 147.

Coggon, A.P. 'London & North-Eastern Railway: The "P.2" Class 2-8-2 Locomotives.' *Journal of the Stephenson Locomotive Society.* March 1945, pp. 36-38.

Coster, P.J. 'From "Cock o' the North" to "Saint Johnstoun."' *Journal of the Stephenson Locomotive Society.* August 1968 pp. 228-248.

Coster, P.J. 'From "Cock o' the North" to "Saint Johnstoun" - Part 2.' *Journal of the Stephenson Locomotive Society.* September 1968 pp. 266-285.

Dobson, K.S. 'Poppet Valve Development on the L.N.E.R.' *The Railway Magazine.* Volume 96, No. 587, March 1950, pp. 197-200.

'Doncaster Railway Exhibition.' *Doncaster Chronicle.* 31 May 1934.

'Dynamometer Tests of "Cock o' the North."' *The Engineer.* Volume 158, July 1934, pp. 16-17.

'Express Passenger Locomotive with Poppet Valves. L.& N.E. Railway.' *The Locomotive Magazine & Railway Carriage & Wagon Review.* Volume 33, No. 421, September 1927, pp. 273-275.

Farmer, J. 'I Helped to Build Mons Meg.' *Steam World.* No. 43, November 1982, pp. 593-597.

'From Eight to Six-Coupled Wheels.' *The Railway Gazette.* Volume 80, March 1944, p. 303.

Geer, H.E. 'Modern Locomotive Super-Heating. - Part I.' *Journal of the Institution of Locomotive Engineers.* Volume 16, Journal No. 75, Paper No. 196, 1926, pp. 419-450.

Geer, H.E. 'Modern Locomotive Super-Heating. - Part II.' *Journal of the Institution of Locomotive Engineers.* Volume 17, Journal No. 78, Paper No. 211, 1927, pp. 79-100.

'Giant Locomotive Visits Dundee.' *Dundee Courier.* 5 June 1934, p. 5.

Gresley, H.N. 'Address by the President.' *Proceedings of the Institution of Mechanical Engineers.* Volume 133, October 1936, pp. 251-265.

Gresley, H.N. 'Inaugural Address.' *Journal of the Institution of Locomotive Engineers.* Volume 12, Journal Nos 31/32, 1918, pp. 199-214.

Gresley, H.N. 'Locomotive Experimental Stations.' *Proceedings of the Institution of Mechanical Engineers.* Volume 121, July 1931, pp. 23-39.

Gresley, H.N. 'Presidential Address.' *Journal of the Institution of Locomotive Engineers.* Volume 17, Journal No. 81, 1927, pp. 558-568.

Gresley, H.N. 'Presidential Address.' *Journal of the Institution of Locomotive Engineers.* Volume 24, Journal No. 121, 1934, pp. 617-625.

Gresley, H.N. 'The Three-Cylinder High-Pressure Locomotive.' *Proceedings of the Institution of Mechanical Engineers.* Volume 109, July 1925, pp. 927-967.

Holmes, V.W. 'A New Infinitely Variable Poppet Valve Gear.' *Journal of the Institution of Locomotive Engineers.* Volume 21, Journal 102, Paper No. 278, 1931, pp. 481-524.

'Improved Type Feed Water Heating Apparatus, L.N.E.R.: Particulars of the New A.C.F.I. Apparatus Fitted to two of the Pacific type Locomotives.' *The Railway Engineer.* Volume 50, November 1929, pp. 442-444.

Langley, Brigadier C.A. and Lieutenant-Colonel G.R.S. Wilson. 'Report on the Derailment which Occurred on the 16th March 1951 at Bridge Junction, Doncaster in the Eastern Region of British Railways.' Ministry of Transport, 1952.

'Latest in Engines Comes North.' *Dundee Courier.* 13 June 1934, p. 10.

'L. And N.E. Railway - Three-Cylinder Eight-Coupled Passenger Locomotive.' *The Engineer.* Volume 158, June 1934, pp. 550-552.

Livesay, E.H. 'André Chapelon and the Steam Locomotive.' *The Engineer.* Volume 200, July 1955, pp. 140-143.

Livesay, E.H. 'Scottish Locomotive Experiences: No. III.-L.N.E.R. Edinburgh and Dundee Trains, "P2" Class Engines.' *The Engineer.* Volume 168, October 1939, pp. 342-344.

Livesay, E.H. 'Scottish Locomotive Experiences: No. IV.-L.N.E.R. Edinburgh-Aberdeen Trains, "V2" Class Engines.' *The Engineer.* Volume 168, October 1939, pp. 366-368.

'L.M.S. & L.N.E. Railways: Locomotive Testing Station.' *Journal of the Stephenson Locomotive Society.* October 1947, p. 215.

'L.N.E.R. Locomotive Conversion.' *The Engineer.* Volume 176, November 1943, p. 410.

'L.N.E.R. Three-Cylinder Eight-Coupled Locomotive.' *The Engineer.* Volume 162, July 1936, p. 20. 'L.N.E.R. 2-8-2 Locomotive "Earl Marischal."' *The Railway Magazine.* Volume 76, February 1935, p. 112.

'Locomotive Conversion on the L.N.E.R.' *Engineering.* Volume 156, November 1943, pp. 427-428.

'Locomotive Feed Water Heater.' *The Engineer.* Volume 48, December 1929, p. 632.

McDermid, W.F. 'The Locomotive Blast-Pipe and Chimney. (Part II).' *Journal of the Institution of Locomotive Engineers.* Volume 23, Journal No. 112, Paper No. 300, 1933, pp. 162-224.

'New 4-6-0 Locomotive for G.E. Section L.N.E.R.' *The Railway Magazine.* Volume 63, Issue No. 377, November 1928, p. 370.

Nock, O.S. 'British Locomotive Practice and Performance.' *The Railway Magazine.* Volume 106, May 1960, pp. 342-350.

Place, P. 'Locomotive Testing Plants (With Special Reference to the Testing Plant at Vitry).' *Journal of the Institution of Locomotive Engineers.* Volume 25, Journal No. 125, Paper No. 338, 1935, pp. 380-415.

Poultney, E.C. 'Poppet Valve Gears as Applied to Locomotives.' [Summary of Lecture given at a Meeting in Leeds] *Journal of the Institution of Locomotive Engineers.* Volume 20, Journal No. 97, 1930, pp. 704-715.

Robson, T. 'The Counter Pressure Brake Method of Testing Locomotives.' *Journal of the Institution of Locomotive Engineers.* Volume 33, Journal No. 173, Paper No. 411, 1933, pp. 171-207.

Sauvage, E. 'Feed-Water Heaters for Locomotives.' *Proceedings of the Institution of Mechanical Engineers.* Volume 103, June 1923, pp. 715-734.

'Solving the Smoke Problem.' *Railway Wonders of the World.* Part 15, May 1935.

Spencer, B. 'The Development of L.N.E.R. Locomotive Design. 1923-1941.' *Journal of the Institution of Locomotive Engineers.* Volume 37, Journal No. 197, Paper No. 465, 1947, pp. 164-226.

'Test of 2-8-2 Type Locomotive No. 2001, London and North Eastern Railway.' *Engineering.* Volume 138, July 1934, p. 20.

'Testing A Locomotive.' *Railway Wonders of the World.* Part 7, March 1935.

'The A.C.F.I. Feed-Water Heater: The Apparatus has been fitted to Locomotives of the 4-4-2 and 4-6-0 Types on the L.N.E.R.' *The Railway Engineer.* Volume 49, July 1928, pp. 267-269.

'Third-Class Sleeping Cars in Great Britain.' *The Railway Magazine.* Volume 63, Issue No. 377, November 1928, pp. 343-350.

'2-8-2 Type Express Passenger Locomotive.' *Journal of the Institution of Locomotive Engineers.* Volume 24, 1934, pp. 468-471.

'2-8-2 Type Three-Cylinder Express Locomotive; L.N.E.R.' *Engineering.* Volume 137, June 1934, pp. 621-623; 715; 728-729.

Webber, A.F. 'The Proportions of Locomotive Boilers.' *Journal of the Institution of Locomotive Engineers.* Volume 27, Journal No. 140, Paper No. 378, 1937, pp. 688-763.

Cock o' the North – Relevant Dates

Early 1932 - Approval received for a new locomotive for Edinburgh-Aberdeen passenger traffic.

March 1932 - First outline drawing.

February 1933 - Approval for construction.

March 1933 - Order no. 330 placed at Doncaster Works.

Late 1933 - Construction begins.

22nd May 1934 - No. 2001 *Cock o' the North* completed at Doncaster Works.

26, 27th May - Doncaster Works exhibition.

1st June 1934 - Displayed at King's Cross and demonstrated to the press.

2nd June 1934 - Present at Ilford exhibition.

4th June 1934 - First journey between Edinburgh and Aberdeen.

5th June 1934 - The locomotive was viewed by the Lord Provost and public at Aberdeen Joint station.

6th June 1934 - Edinburgh Waverley station hosts *Cock o' the North*.

11-16th June 1934 - Working from King's Cross.

19th June 1934 - First dynamometer car trial.

21st June 1934 - No. 2001 begins working from Doncaster.

27th June 1934 - Second dynamometer car test.

2nd, 3rd, 5th July 1934 - Indicating shelter fitted to the engine for testing the performance of the cylinders.

13th - 19th July 1934 - Trials carried out to set the Kylchap blastpipe and chimney.

20th - 30th July 1934 - 'Sine wave' superheater elements removed and other minor modifications.

31st July 1934 - Arrives in Scotland and allocated to Edinburgh Haymarket shed.

1st, 2nd August 1934 - Scottish Area test the locomotive.

24th - 27th August 1934 - Minor modifications at Doncaster Works.

1st - 5th October 1934 - Coal Consumption tests in Scotland.

23rd October - 1st November 1934 - Trials with different blastpipe and chimney settings.

4th December 1934 - Departs from Doncaster Works for Vitry-sur-Seine testing station.

8th December 1934 - *Cock o' the North* runs on the locomotive table for the first time.

13th December 1934 - The tests begin, but are halted after an axlebox overheats.

17 December 1934 - Trials are resumed with no. 2001.

1st - 6th February 1935 - *Cock o' the North*'s road tests take place.

12th February 1935 - Trials in France concluded.

17th February 1935 - No. 2001 exhibited at Gare du Nord, Paris.

21st February 1935 - *Cock o' the North* returns to England.

April/May 1935 - The locomotive works from Doncaster shed.

4th, 5th May 1935 - Present at Stratford Works exhibition.

June 1935 - No. 2001 begins regular service in Scotland from Edinburgh Haymarket depot.

16th, 17th May 1936 - Kirkcaldy exhibition.

September 1937 - April 1938 - *Cock o' the North* rebuilt with piston valves, streamlined front and exhaust steam injector.

June - September 1944 - Rebuilt with a Pacific wheel arrangement.

October 1944 - Allocated to Aberdeen Ferryhill shed.

19th June 1946 - Appeared at Edinburgh exhibition.

August 1946 - Became no. 501.

May 1948 - Renumbered 60501.

September 1949 - Allocated to Edinburgh Haymarket depot.

November 1949 - Transferred to York shed.

November - December 1950 - *Cock o' the North* spends a month on loan to Neville Hill shed, Leeds.

16th March 1951 - Involved in an accident at Bridge Junction, Doncaster.

October 1952 - Returns to traffic after a general repair with a diagram 118 boiler.

February 1960 - Withdrawn from traffic and subsequently scrapped at Doncaster Works.